Students

ATE OF ISSUE

D1345895

AN INTRODUCTION TO BIORHEOLOGY

AN INTRODUCTION
TO BIORHEOLOGY

by

G. W. SCOTT BLAIR, M.A. (Oxon), Ph.D., D.Sc. (Lond.), F.R.I.C., F.Inst.P.

Chapter XII on Botanical Aspects by

D. C. SPANNER, Ph.D., D.Sc. (Lond.), F.Inst.Biol.

Foreword by Professor A.L. Copley, M.D. (hon. caus.) (Heidelberg)

ELSEVIER SCIENTIFIC PUBLISHING COMPANY
AMSTERDAM - OXFORD - NEW YORK 1974

ELSEVIER SCIENTIFIC PUBLISHING COMPANY
335 Jan van Galenstraat
P.O. Box 211, Amsterdam, The Netherlands

AMERICAN ELSEVIER PUBLISHING COMPANY, INC.
52 Vanderbilt Avenue
New York, New York 10017

Library of Congress Card Number: 73—85229

ISBN 0-444-41160-7

With 51 illustrations and 1 table

Copyright © 1974 by Elsevier Scientific Publishing Company, Amsterdam

Printed in The Netherlands

FOREWORD

I feel greatly honored to have been asked by the author to write a Foreword for this book.

As the title indicates, this book gives an introduction to the science of biorheology. The author has spared himself no pains in his endeavor to present the main fields of this rapidly growing science in simple, intelligible and most inviting ways. The readers who are not familiar with biorheology and those who are actively engaged in advancing knowledge in any one of its fields, I believe, will be delighted about the author's presentation.

George W. Scott Blair is endowed with several gifts, rare among scientists. His originality as experimentalist and thinker is combined with that of an enthusiastic teacher. He captivates audiences by the unique presentation of highly intricate subjects with eloquence and an uncommon capacity to make them easily understood. He appears to humanize both the subject he is talking about and the audience by his warmth and wits, thus providing a different dimension in addition to the information he conveys in what he sets out to do from his vast knowledge, critical appraisal and deep insight. As George W. Scott Blair entices a listening audience, he is a master of keeping alive the interest of the readers of the introductions he wrote in his books and articles on the physical science of rheology. His "Introduction to Biorheology" is such an admirable text.

No one other than the author could have given a more stimulating account of this newly organized Science and no one has worked in more fields of biorheology than Scott Blair. He emphasized in his Preface that this book is not meant to be a treatise, but an attempt to provide an overall picture of the scope of the subject with emphasis on the practical rather than the purely theoretical aspects.

George W. Scott Blair received from his peers many honors for his high scientific attainments. The Poiseuille Gold Medal was awarded to him at the 2nd International Conference on Hemorheology, held in 1969 at the University of Heidelberg, for his outstanding contributions to hemorheology and other fields of biorheology. The Award is the highest distinction which the International Society of Hemorheology (now named the International Society of Biorheology) bestows to honor the accomplishments of the biorheologist so chosen.

In 1948 I had the good fortune to make the author's personal acquaintance during the 1st International Congress on Rheology at Scheveningen, The Netherlands. The year, during which the Congress was held, marked the beginning of the new science of biorheology as that of our friendship. Five years earlier Scott Blair invited me to chair in Oxford a conference on Rheology and Medicine, which, because of the war, I was unable to accept. During my stay in Paris, where I worked for five years until 1957, we visited each other on several occasions in England and in France. Thereafter, for nearly three years of my residence in London, before returning to New York, I had the great privilege and pleasure to do experimental work with him in hemorheology. We organized together the Meeting on "The Flow of Blood in Relation to the Vessel Wall", held in 1958 at Charing Cross Hospital Medical School of the University of London.

In 1959 we worked together on the organizing committee of the Symposium on "Flow Properties of Blood and Other Biological Systems", held at the Physiology Laboratory of Oxford University and convened jointly by the Faraday Society and The British Society of Rheology. In 1962 we founded, encouraged by Mr. Robert Maxwell, the international journal, "Biorheology", published quarterly by Pergamon Press, Oxford. It is now the Official Journal of the International Society of Biorheology which last year became an Affiliated Commission of the International Union of Pure and Applied Biophysics. As Editors-in-Chief of "Biorheology", which, beginning in 1974, will be published bimonthly, we have met regularly each year since its foundation, and, jointly also in this activity, did our share towards the advancement of this young and vigorous science.

As psychology may be regarded as a branch of biology, the field of psychorheology, a term proposed by G. W. Scott Blair and F. M. V. Coppen, belongs to biorheology. It is therefore appropriate that this book includes an appraisal of findings in studies of the relation between the assessing of rheological properties by the handling of materials and their measurements by means of physical apparatus.

In many fields of botany, biorheological phenomena play a great role and in some areas significant contributions have been made to both the physical science of rheology and the biological sciences. It is therefore an asset that a chapter on botanical apescts, written by Prof. D. C. Spanner, is included in the book.

"An Introduction to Biorheology" represents the first book of its kind. I am convinced that the book, which the author dedicated to the medical profession, will be also of value to biologists and to scientists of all descriptions. There has been a need for such a book which now has been met by George W. Scott Blair's "Introduction to Biorheology". It makes the readers aware of the importance of biorheology to the medical and biological sciences and its applications to the practice of medicine and surgery. I believe that this book will serve also to stimulate interest towards further progress in biorheology for the better understanding of nature and for the benefit of the health of man.

New York, N. Y. (U.S.A.),   *A. L. Copley*
March 20, 1973

PREFACE

I introduced my second book* with a quotation from Epicurus: "Vain is the word of a philosopher which does not heal any suffering of man." When I wrote this book, rheology was almost exclusively concerned with industry: only a few attempts had been made to apply this branch of physics to medicine and its ancillary subjects.

When I decided to write the present book, about a year ago, I had no idea that the subject of Biorheology would develop at such a remarkable speed while I was writing. This is not only a result of the holding of the 1st International Congress on Biorheology in Lyon in September 1972, because much of the work now being published must have been well started before that date. The journal "Biorheology", of which (with Professor A. L. Copley) I am Co-Editor-in-Chief, has been published now for ten years as a comparatively small journal, but, apart from the Proceedings of the Congress, which will be published in it, the number of papers submitted during the last few months has greatly increased. It seems that physiologists and doctors have come, very recently, to realize the potentialities of rheology to help them in their work.

This has placed me in something of a dilemma: it seemed best to continue with the writing of the book, because, so far as I know, no other book covering the field of biorheology exists (other than Proceedings of Congresses and a series of essays on specific subjects published in 1952)**.

Through the courtesy of the publishers, I am to be allowed to add a brief Appendix—a kind of Stop Press, referring to the most important work coming to my notice after the main part of the book has been printed.

I would stress that this book is essentially "an *introduction*": it is not a treatise and it makes no claim at all to cover the entire subject of biorheology. A number of books have been published recently on that aspect of the subject which has, not unnaturally, included a high proportion of published work so

---

* Scott Blair, G. W., *A Survey on General and Applied Rheology* (Sir Isaac Pitman and Co., London, 1st edn., 1943).
** Frey-Wyssling, A., *Deformation and Flow of Biological Systems* (North-Holland Publ. Co., Amsterdam, 1952).

far; i.e. haemorheology: the study of blood, its components and vessels. One of these books alone contains about a thousand references. A book on protoplasm is also almost equally lavishly documented.

Rather, what I have attempted to do is to provide an overall picture of the scope of the subject as I see it, tending to stress some of the lesser-known aspects and, in the chapters concerned with blood and protoplasm, to give a somewhat sketchy account of the literature, emphasizing as far as possible, those aspects that would seem to have been given less space by the specialized authors. Perhaps I have over-emphasized my own experimental work. If so, this is because I am most familiar with it. I have also tried to stress the practical rather than the purely theoretical aspects.

I have listed a number of books for further reading and have given a goodly number of references to original papers, some 400 in all, throughout the book.

In writing a technical book, one must first think carefully for whom it is intended. This book is not meant for rheologists, but rather for biologists, physiologists and especially medical men who may be wondering what "all this rheology" is about and whether it could help them in their work.

I dedicate it to the medical profession, among the hardest worked and devoted of all professions, and especially to those doctors and surgeons whose skill and care have made it possible for me to write a book at the age of seventy; and to those with whom I have had the privilege of working, on and off, over many years (though, alas, quite a few are no longer with us).

It has been difficult to know just what to include under the heading of Biorheology. I was particularly uncertain about a chapter on the psycho-physical aspects, but was encouraged to include this, both by the publishers and by Professor A. L. Copley.

I have introduced a minimum of mathematics in this book. Of course any-one seriously taking up research in biorheology must be familiar with general rheological theory; but so much research these days is done, quite rightly, by small groups of workers and in a team of, say, a doctor, a physiologist and a rheologist. Perhaps only the last need worry too much about the mathe-matics; although in a good team, each member will learn all he possibly can about the specialities of his colleagues.

Although I have had to introduce (and explain) quite a number of rheo-logical terms, I have tried to avoid an excess of nomenclature; nor have I insisted on using one symbol (such as Greek tau) for only one purpose throughout the whole book. Indeed rheologists are getting so short of symbols, even allowing for superiors, inferiors and different founts, that the Latin and Greek alphabets are proving inadequate. (Unfortunately the Cyrillic alphabet is of little use, since so many of the printed letters are identical with, or very

similar to, the Latin.) In his latest book* Professor Reiner has had to introduce several Hebrew letters. Of course, in such a book, every symbol must
have a unique significance, but this seemed hardly necessary in the present
work.

Many of my friends have helped me, especially with biological nomenclature. I was trained as a chemist, and although I have specialized in rheology
now for some forty-five years, "biorheology", except in very recent years, has
always been for me a fascinating side-line, and my knowledge of biological
terminology is limited.

I knew that I could not possibly tackle a chapter on botany, and was very
fortunate in persuading Prof. D. C. Spanner to write this for me.

I am also very grateful to my old friend Professor A. L. Copley, of New
York, not only for reading through my manuscript, but also for contributing
a foreword.

My first book** had a foreword by Professor E. C. Bingham, "founder of
modern rheology" who gave the science its name. Professor Copley, a leading
biorheologist, first named Biorheology in 1948, was the first President of the
International Society of Biorheology, and, at the recent Congress, he was
awarded the Poiseuille Gold Medal of that Society "for his outstanding
contributions, both in the field of research and organization, to the field of
Biorheology". Following the Congress, he received the further distinction of
an Honorary Doctorate in Medicine at the University of Heidelberg.

I had the privilege of working with him when he was in London some years
ago and owe much to his encouragement and help.

My best thanks are due to Mrs. D. Steer, who converted my somewhat
illegible scrawl into neat typing. I am much indebted to my wife and to my two
friends, Erika Berkenau and Dr. Graham Patrick for completing the indexing
and for work on the proofs.

Grist Cottage, Iffley, Oxford (England),   *G. W. Scott Blair*
January 1973

* Reiner, M., *Advanced Rheology* (H. K. Lewis and Co., London, 1971).
** Scott Blair, G. W., *Introduction to Industrial Rheology* (J. and A. Churchill, London,
1st edn., 1938).

CONTENTS

Chapter I

DEFINITIONS AND LIMITATIONS OF RHEOLOGY AND BIORHEOLOGY; ELECTRORHEOLOGY

The field of scientific studies is so wide that it is convenient to divide it into separate disciplines. Thus we have Physics, Chemistry, Biology, Psychology etc., and subdivisions such as Botany, Geology, Paleontology etc.

But many of the most interesting fields of work cover at least two of these major categories; in which case it is customary to link the two names together, and we have Physicochemistry (or Physical Chemistry), Biochemistry, Paleobotany etc., sometimes written as two words, sometimes hyphenated and sometimes as a single word.

It is evident that Biorheology covers a combined study of Biology and Rheology. The meaning of the former term is well-known: the Shorter Oxford Dictionary (3rd edn.) gives, for the modern use: "The science of physical life, dealing with organized beings, or animals or plants, their morphology, physiology, origin and distribution."

The term Rheology, being newer, is less well-known. The same Dictionary gives no definition: only some rarely used combinations of "rheo" such as "rheometer" as applied to electrical measurements.

The German form of the word (Rheologie) and the description of a small viscometer as a "Microrheometer" are to be found at early dates; but, in its modern sense, the term Rheology was coined by Prof. E. C. Bingham and formally adopted and defined at the Foundation Meeting of the American Society of Rheology in December 1929 in Washington, D.C. as "The science of the deformation and flow of matter".

This definition, with occasional slight modifications, and with the term appropriately spelt in different languages, is now universally accepted.

However, it has also been generally accepted that studies which are not concerned with the structure or composition of materials, such as hydrodynamics and aerodynamics, are not included in rheology.

All materials deform and/or flow (though this is not really what Heraclitus meant when he said "panta rhei"), so that the range of rheology is very wide. Rheologists are perhaps of most obvious practical use in studying materials whose composition and structure is exactly defined, so that the widest field of rheology has always been, and probably always will be, concerned with synthetic high polymers ("plastics").

However, biological systems were studied long before a separate study of "biorheology" was recognized—for example, Poiseuille's classical work on blood, which will be discussed later in this book; and there will be other examples. But it was not until 1948 that, at the 1st International Congress on Rheology, held at Scheveningen, The Netherlands, that Dr. (now Prof.) A. L. Copley proposed the use of the term Biorheology. The same limitations have been applied as in other branches of rheology: thus work on the structural properties (elastic moduli etc.) of wood would not be considered as "biorheology" even though wood has a biological origin; whereas a study of the flow of sap through trees definitely is (see Chapter XII), because it is concerned with biological phenomena.

The Society founded in Washington in 1929 was originally intended to be international, but this was never effected and it is strange that, although very many countries now have Rheological Societies or Groups, there is still no "International Society of Rheology"; only an *ad hoc* International Committee to decide on the dates and locations of International Congresses etc., of which six have now been held*.

At a Conference at the University of Iceland (Reykjavik) in 1966, an "International Society of Haemorheology**" was founded, the name again being proposed by Prof. Copley, to study the rheological properties of blood, its components and of blood vessels. At the 2nd Congress (Heidelberg, 1969), however, the name and scope of the Society were changed to include all Biorheology; so that there is now the rather strange situation that the biorheological branch of this subject has its own International Society, whereas the whole of rheology does not.

The journal "Biorheology" (Pergamon Press, Oxford) started publication in 1962 and later became the Official Organ of the International Society, which is itself affiliated to the International Union of Pure and Applied Biophysics. (Of course, many biorheological papers are published in other journals.) Thus we have:

---

* The author has just been informed that the Committee, reconstituted in 1968, now also deals with nomenclature, etc. and the encouragement of the formation of new National Societies, as well as making contact with other International Bodies.
** In U.S.A. spelt "Hemorheology".

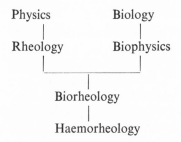

Before proceeding further, it would be well to consider briefly the fundamental principles of rheology; "briefly", since there are already many textbooks on the subject. A few are listed at the end of this chapter.

The two fundamental laws of elastic deformation and of flow respectively were both "discovered" in the second half of the 17th century. "Discovered" is perhaps not quite the right word, because, whereas Hooke found his law as a result of experiments on the stretching of long metal wires and of coiled springs, Newton proposed his as being the simplest relation that might be supposed to exist between what we now call stress and rate of shear or flow, and he was careful to add "other things being equal".

Hooke found that, for the very small recoverable deformations that a metal wire can undergo without either "flowing" or breaking, and also for the large extensions that can be produced in a coiled spring without stretching the metal itself very much, the extension is proportional to the applied force.

Newton proposed "that the resistance which arises from the lack of slipperiness [modern: "viscosity"] of the parts of the liquid, other things being equal, is proportional to the velocity with which the parts of the liquid are separated from one another".

Following the work of Galileo (although some of his "experiments" were probably only "thought experiments"), the scientific climate at the time of Hooke and Newton strongly favoured the idea of the *simplicity* of Nature. If only they could be found, natural laws were believed to be essentially simple.

Although mistaken, this idea was profitable at the time, because it is important to know the basic laws of gravitation, deformation, flow etc., which hold under many normal situations or for many materials before going on to study the exceedingly complex modifications that are sometimes needed in practice. Perhaps it is just as well that Euclid's geometry was believed to have universal validity for so long before other geometries were attempted. It so happens that, because atoms cannot be pulled far apart without permanent separation, Hooke's law holds for many materials. "High-elasticity" materials like rubber and many plastics were hardly known, or unknown in Hooke's day. Many liquids such as water, alcohol and all but the heaviest oils obey Newton's law to a high degree of accuracy. When we discuss blood in a later chapter, we shall see that there has been much sterile argument as to whether its liquid components are, or are not, Newtonian. In nature, nothing is perfect: the proper question to ask is whether, under certain experimental conditions, deviations from the simple equation are significant.

Some elastic materials show a more complex (non-linear) relation between stress and strain and others, though they recover their original dimensions when the stress is removed, do so only slowly; i.e. there is a viscous resistance delaying the elastic recovery. Many suspensions and emulsions (solid and liquid particles dispersed in a liquid medium) show a fall in viscosity as the stress is increased; some show a rise. In some cases, the fall in viscosity (or consistency) is recovered slowly on standing ("thixotropy") and a few rheologists, pointing out that "slowly" is an imprecise term, would use the term "thixotropy" to describe all recoverable losses in consistency produced by shearing.

When the structure, broken down by shearing, is recovered so fast that all that is observed is a fall in viscosity with increasing rate of shear, the original value being re-established when the shear rate is lowered, the material is said to show "shear thinning". The reverse effect, i.e. an increase in viscosity with increasing shear rate, is called "shear thickening". This may be caused by an expansion of the system such that closely packed particles have to move apart into open packing before they can pass one another. Sand and water show this phenomenon very clearly. This is called "dilatancy" but this term is often loosely used to describe any type of thickening, even when there is no evidence that the system "dilates". Most rheologists deprecate this use of the word.

In this book, we will not attempt to give the long list of technical terms used to describe the many deviations from simple deformation and flow. Reiner and Scott Blair [1] have attempted a complete list and, studying this and various other lists prepared mainly in The Netherlands, Australia and France, a Committee of the British Standards Institute has worked for some years on the preparation of what it is hoped will be a generally accepted list of terms, in English. (Reiner and Scott Blair included translations in French, German and Russian.)

Apart from "thixotropy", there is one other term which is liable to confuse biorheologists and which should therefore be mentioned here.

In 1936, two pupils of Freundlich (who invented the term "thixotropy"), Juliusburger and Pirquet [2], found that certain *thixotropic* systems would reset more quickly if the vessel containing them was slowly rotated. They called this phenomenon "rheopexy".

Some materials increase their consistency when sheared and "soften" when left to stand. This phenomenon was for many years very sensibly called "negative or inverse thixotropy". Unfortunately, some years ago, certain workers who had not troubled to read Juliusburger and Pirquet's paper with care, started to call this phenomenon "rheopexy". So many others have now blindly followed that, if a biorheologist reads that some material is "rheopectic" he can only guess, from the date of the paper he is reading, to which phenomenon the term applies.

The present author has been accused of being "obsessed by nomenclature". He is not ashamed: nomenclature is extremely important if rheologists are to understand one another. But the present book will not be overburdened with this question.

Unfortunately, biological systems are generally extremely complex rheologically, and the reader who is seriously interested in biorheology should read at least one elementary textbook on general rheology before attempting a serious study of biological systems.

But two more very important fields of rheology must be briefly discussed here before proceeding to consider biological systems.

It has already been mentioned that rubber-like materials have become widely known only in recent times. It used to be thought that their elasticity was like that of a coiled metal spring. When the spring is stretched, the energy is held within the system: strong attractive forces between the atoms try to bring them together again. However, it is now known that in rubber and many other "high-elastic" materials, the mechanism is quite different.

If we imagine a long piece of string being bombarded from all sides by large numbers of small particles, it is clear that all sorts of configurations will be formed. It is rare that the string will be fairly straight. There are far more possible configurations in which it will be coiled up. If we apply a force and straighten the string and then let go, it will not be long before the bombardment to which it is subjected will cause it to coil up. The heavier the bombardment (i.e. the higher the temperature) the faster this process will go. This is exactly what happens with rubber-like materials. As we shall see in a later chapter, in the case of muscle, the situation is greatly complicated by electro- and physicochemical changes.

The other question which is bound to affect those studying biorheology (or almost any rheology) for the first time is the rheologists' method of dealing

6

with systems which are both viscous and elastic (briefly referred to above). Strictly speaking, one should distinguish two cases: The first would be represented in its simplest form by a steel spring attached in series to a piston sliding in a cylinder of oil. The latter, used by engineers to "damp" the vibration of some machinery, is called a "dashpot". Such systems, and much more complicated systems like them, are essentially "viscous", i.e. their deformation is not fully recovered when the force is removed. Strictly speaking, such systems are called "elasticoviscous". In others, in which the spring is attached in parallel to the dashpot, the recovery is complete but slow. These are strictly called "visco-elastic". But in fact many rheologists use the latter term for both.

In earlier times, rheologists made (or drew) "models" consisting often, of many springs and dashpots, and also "sliders" (a weight resting on a flat surface) as well as other devices to describe the behaviour of complex systems.

But all these "models" can be described by quite simple mathematical equations and it is characteristic of the climate of modern physics that the equations are now generally used instead of the models.

Before closing this chapter, a very brief description should be given of the principal factors that are measured by rheologists and, in Chapter II, some experimental methods will be discussed.

English is a rich language, so we are fortunate in having many synonyms. The physicist can arbitrarily define each synonym to have different scientific meanings. Thus, in ordinary parlance, the words "stress" and "strain" are synonymous. The rheologist, however, uses the former term to refer to a force or system of forces and the latter, purely geometrically, to mean changes of shape and/or volume.

It is as useless to ask whether forces produce changes of shape, or whether changing the shape produces the forces as it is to ask which came first: the hen or the egg. It is easiest to start by imagining a small cube of some material on which forces are exerted in all directions. Imagine this cube standing on a table and consider the force acting on to the top surface at right angles to the table. According to Newton's Law, there will be an equal and opposite force by which the surface of the table resists this compression. This means that, although the cube has six sides, we need worry only about three of them. These are usually called x, y and z. If we consider any force acting on the x-face this may be expressed as made up of three parts: (1) a component at right angles to the x-face (called "normal", which we may write as p_{xx} or simply p_x); (2) a component across the x-face in the y-direction which we write p_{xy}; and (3) a component across the x-face in the z-direction, p_{xz}.

The whole "state of stress" could thus be expressed in terms of nine magnitudes—but it can be shown that symmetrical pairs, such as p_{xy} and p_{yx} are identical, so that we have six independent components. This group of six components, taken together, constitutes what is called the "stress tensor".

Similarly, there is a corresponding group of strain components (strain being taken as a *relative* deformation; i.e. if the cube were twice as tall, to get the same strain we should need twice the deformation). This gives six strain components, constituting the "strain tensor". Likewise there is a rate-of-strain tensor.

Normal liquids have the same properties in all directions, i.e. are *isotropic*, so they have only one characteristic property, the viscosity, which is defined as the stress divided by the rate of change of strain (or rate of shear). Solids may have different properties in different directions (anisotropic)—consider, for example, wood along and across the grain.

Since elastic moduli are defined as the ratios of the stress to the strain components, and there are six of each, one might suppose that there would be 36 possible elastic moduli (or their reciprocals, called "compliances"). In fact again symmetrical pairs are equal, so the greatest possible number is reduced to 21. Wood, which has the same properties in two of the three planes (across the grain) has 9 separate constants.

The term "stress" is sometimes loosely used to refer to the separate components of the stress tensor (forces per unit area). Components which are tangential (not "normal") produce "shears".

The professional rheologist will find a few over-simplifications in the above short description: my intention is to give those new to biorheology some idea of what the rheologist means by the terms he most commonly uses. For more precise definitions, he should consult the textbooks quoted at the end of this chapter.

Rheological behaviour in relation to electrical phenomena

There is one aspect of biorheology, not dealt with in the subsequent chapters of this book, but which may play an important part in future biorheological research: this is the relation between rheological properties and electrical phenomena. The subject is well summarized in a review article by Fukada [3]. Electrical polarization plays so important a part in the physiology of all living matter that it is surprising that its connection with rheological behaviour would appear to have been largely overlooked. Almost certainly it will form an important field in future research.

Elastic deformation under mechanical stress and polarization under an

8

electric field are closely analogous and are known to be linked in the case of synthetic polymers.

There was also, at one time, considerable controversy over the effect of electric fields on the viscosity of simple liquids. Sokolov and Sosinskiĭ [4], in the U.S.S.R., claimed an increase in the viscosity of organic liquids under the influence of both alternating and direct electric fields. Andrade [5] and Andrade and Dodd [6] came to somewhat different conclusions, distinguishing between the behaviour of polar and non-polar liquids and even subdividing the former into two separate classes (see also Andrade and Hart [7]).

In his earliest paper, Andrade found an increase in the viscosity of certain liquids under the influence of an electric field: in the later work, he found no effect for non-polar liquids, but that an electric field transverse to the line of flow, increased the viscosity of certain polar liquids. He ascribed this to an accumulation of ions around the electrodes. Using alternating currents, the effects varied with the frequency. There were considerable differences of opinion between the British and the Russian workers. So far as the author is aware, this complex subject has never been fully clarified*.

Whether such phenomena have any bearing on biological systems is problematic; but the "piezo-electric effect", as discussed by Fukada [3], almost certainly does. Elastic strain caused by applying an electric field is called the "converse piezo-electric effect" and electric polarization produced by mechanical stress is called the "direct piezo-electric effect". The quantitative equations relating the electrical and elastic constants will not be given here: the original papers should be consulted. We have already seen that the stress tensor has six components: the polarization has three, i.e. those in the x-, y- and z-directions. Combining these, the piezo-electric modulus has a maximum of eighteen components.

The Japanese workers have found these effects in a number of biological materials, especially bone and tendon which have seven finite components. A sine curve is shown in Fukada's paper [3] relating the angle between the direction of the force and the axis of the tendon to the piezo-electric modulus; for an Achilles tendon. A very simple formula gives the anisotropy.

The effect is ascribed to the uniaxial orientation of collagen crystallites and the piezo-electric effects inherent in these crystals.

Very good straight lines are shown relating pressure to polarization in the direct piezo-electric effect on bone and also between the electric field and the mechanical strain in the converse piezo-electric effect.

Shamos and Lavine [8] studied the piezo-electric effect in skin and callus

---

* See APPENDIX, p. 199

and found an anisotropy similar to that in bone. Fukada himself studied blood vessels, intestines and trachea dehydrated by soaking in alcohol for some weeks. The piezo-electric modulus is naturally much smaller than that in bone or tendon, but the anisotropy is clearly seen. A comparison of data from the aorta and from the veins suggests that, for the veins, the orientation of protein fibres in the natural state was in the lateral direction, in contrast to the aorta.

Other Japanese workers found that a weak electric current accelerates the formation of bone and Bassett et al. [9] showed that a direct current applied continuously to a dog's thigh caused an appreciable formation of bone. These effects should be followed up and might lead to valuable suggestions for orthopaedists. The effects of electrical potentials applied to the ear are already well-known.

The charge distribution on a femur subjected to pressure has been examined as a possible approach to in vivo experiments.

It has also been found that a complex calcium phosphate, hydroxyapatite, generates a voltage when a bone, in whose collagen fibres its crystals occur, is bent. The phosphate dissolves when the bone is inactive but redeposits when it is under stress.*

There would seem to be very considerable unexplored possibilities which might be of use to medicine in this little-known field of electrorheology. The work is described here because it does not seem to have been appreciated by workers in the general field of biorheology to be discussed in later chapters.

REFERENCES

[1] Reiner, M., and Scott Blair, G. W., in *Rheology: Theory and Applications* (Ed. F. R. Eirich), Vol. IV, Chapter IX (Academic Press, New York, 1967).

[2] Juliusburger, F., and Pirquet, A., *Trans. Faraday Soc.*, 32: 445 (1936).

[3] Fukada, E., *Biorheology*, 5: 199 (1968).

[4] Sokolov, P. and Sosinskiĭ, S., *Acta physiochim. U.S.S.R.*, 5: 691 (1936).

[5] Andrade, E. N. da C., *Proc. phys. Soc.*, 52: 748 (1940).

[6] Andrade, E. N. da C. and Dodd, C., *Nature*, 143: 27 (1939).

[7] Andrade, E. N. da C. and Hart, J., *Proc. roy. Soc. A*, 225: 463 (1954).

[8] Shamos, M. H. and Lavine, L. S., *Nature*, 213: 267 (1967).

[9] Bassett, C. A. L., Pawluk, R. J. and Becker, R. O., *Nature*, 204: 652 (1964).

BOOKS ON GENERAL RHEOLOGY FOR FURTHER READING

Very elementary: Scott Blair, G. W., *Elementary Rheology* (Academic Press, London, 1969).

*See APPENDIX p. 199

10

Fairly elementary: Reiner, M., *Deformation, Strain and Flow* (H. K. Lewis, London, revised edn., 1960).

Intermediate: Reiner, M., *Lectures on Theoretical Rheology* (North-Holland Publ. Co., Amsterdam, 3rd edn., 1960).

Advanced: Reiner, M., *Advanced Rheology* (H. K. Lewis, London, 1971).

Chapter II

SOME TESTING METHODS SUITABLE
FOR BIORHEOLOGISTS

The principal "properties" measured by biorheologists are viscosities (simple or complex) in which deformations are not recovered on release of stress; elastic moduli, measuring, as we have seen, the relation between recoverable deformations and stresses; and conditions of rupture.

The experimental techniques for measuring such properties do not differ essentially from those used in other branches of rheology, except perhaps in that it is seldom possible to get large samples of biological materials. Most of the rheometers available on the market are designed for quite large samples: in industry, there is generally no shortage of material for testing.

As a result of this, although the *principles* used in biorheological testing are perfectly standard, most workers in this field have had to design or make their own apparatus. Sometimes a commercially available instrument can be suitably adapted but, if any workshop facilities are available, it is generally better to start from scratch.

That this has in fact been the usual practice is easily seen in looking through numbers of such journals as "Biorheology".

Another fairly characteristic factor in many, but by no means all, biorheological studies is that the samples are not homogeneous. A sample of paint, flour dough or a bar of metal will be reasonably homogeneous "in the large"; whereas cervical mucus, bronchial mucus or synovial fluid are grossly heterogeneous. The rheologist is then faced with the difficult problem of deciding how much to homogenize, or separate his sample. Without any homogenization or separation into components, measurements are apt to be highly irregular or meaningless, whereas too much mixing may well destroy the very structure that the rheologist wants to measure. For example, semen, as we shall see in Chapter VII, especially in some animals such as the boar, there are large changes in the rheology of the material during the course of ejaculation. Even in taking a blood sample, changes in composition occur during the process of filling the hypodermic and subsequent mixing must be done with care. Nor can one be sure that the resulting "mixture" is truly representative of the blood as it was in the vein. (Problems of extracorporeal coagulation will be discussed in Chapter VI.)

In this chapter, only a few principal methods of testing will be discussed in some detail and formulae for calculating rheological parameters will be given; but the calculation of the equations will not be included. These are to be found in the standard rheological textbooks (see especially those listed at the end of this chapter).

The most widely studied biological material has certainly been blood and, since this flows in "tubes" in the body, the capillary* viscometer, in one form or another, has been the most popular. This tradition goes back to the classical work of Poiseuille, who, independently of Hagen, first found the general law relating to viscosity (η) to applied pressure (P), length of capillary (L) and radius (R):

$$\eta = \frac{P \pi R^4}{8 L V} \qquad \qquad \text{(Equn. II-1)}$$

where V is the volume of flow per second. (Pressure must be given in dynes. cm^{-2}, i.e. (force/area)$\times g$, g being the acceleration of gravity.) (See Chapter V.)

Viscosity is thus defined as a stress (dimensions ML^{-1}T^{-2})** divided by a shear rate (T^{-1}), hence it has dimensions ML^{-1}T^{-1}. Taking the unit of mass as 1 gram, length 1 cm and time 1 second, the unit of viscosity is called "1 poise" (sometimes written "P" and sometimes "p"). The hundredth part of this unit, the centipoise, "cp", happens to be the viscosity of water at a convenient laboratory temperature. These abbreviations will be used throughout this book, although a more modern system of nomenclature (the "S.I. System") is now being proposed—not without criticism from some physicists.

In more recent times, the capillary viscometer has been much criticized. Many of the standard types, such as the Ostwald (shown in Fig. II-1) operate at a changing head and are therefore unsuitable unless the material may be treated as Newtonian, i.e. obeying Equn. II-1. For certain purposes only this is near enough for blood. Even so, for really accurate determinations of viscosity, corrections should be made for anomalies at the ends of the capillary, including the "kinetic energy correction" to allow for the effects of accelerations with the sudden change of the radius of the vessel through which the material is flowing. But, in work on blood and other body fluids,

---

* The term "capillary" must be used with care. To the rheologist it means simply a long narrow tube, but to the haematologist, it refers only to the very small vessels. Thus the former would call quite a large blood vessel a "capillary", whereas the latter would not.
** M, mass; L, length; T, time.

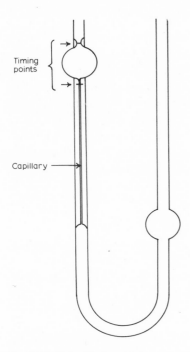

Fig. II (1). Ostwald viscometer.

it is seldom possible to attain an accuracy high enough for these corrections to be needed.

It is, of course, easy to design a capillary viscometer in which the head remains adequately constant: e.g. by having a horizontal capillary attached at each end to wide tubes bent into a vertical position. If we measure the rate of fall of the liquid in the vertical arm, the equation for the kinematic viscosity (v) (ignoring a usually small surface tension correction) is:

$$\frac{\eta}{\rho} = v = \frac{t}{\ln (h_0/h)} \cdot \frac{g R_c^4}{4 R_w^2 L} \qquad \text{(Equn. II-2)}$$

where ρ is the density; when h_0 is the initial head, h is head at time t, R_c and R_w are the radii of the capillary and wide tubes respectively and L is the length of the capillary. The vertical tubes can, if necessary, be themselves attached to wider containers in which the level is kept constant. But there is a further difficulty. The rate of shear (and consequently the stress, since viscosity is given by their ratio) varies from zero at the wall of the capillary to a maximum value at the centre. If experiments are done over a wide range of externally applied pressures, this may not matter much; but for accurate work to study the effect of varying shear rate on viscosity, it is a definite disadvantage.

Many forms of microcapillary viscometer have been used. With small samples, it is often an advantage to measure the rate of filling or emptying of a capillary. In these cases, the length of the column of material (l) varies. The equations for (a) filling and (b) emptying are as follows:

(a) $\quad \eta = \dfrac{t}{l^2} \cdot \dfrac{PR^2}{4}$ (Equn. II-3)

when t is the time.

(b) $\quad \eta = \dfrac{t}{l_0^2 - l^2} \cdot \dfrac{PR^2}{4}$ (Equn. II-4)

where l_0 is the initial length of the column. Very short columns are subject to errors caused by surface tension.

A nearer approximation to a constant shear rate may be obtained by means of two coaxial cylinders. The invention of this method is generally ascribed to Couette. The outer cylinder (radius R_2), which contains the sample, is generally rotated at a constant angular velocity Ω. The inner cylinder (radius R_1) is suspended on a torsion wire centrally within the outer cylinder, so that the material to be tested fills the annular space between the cylinders. The viscosity of the liquid causes a torque to be transmitted to the inner cylinder which can be measured, for example, by attaching a small mirror to the torsion wire and recording the deflection of a beam of light reflected from the mirror.

The equation for the viscosity of a Newtonian liquid is slightly complicated (known as the Margules equation):

$$\eta = \frac{M (R_2^2 - R_1^2)}{4 \pi \Omega h R_1^2 R_2^2}$$ (Equn. II-5)

where M is the measured torque, and h is the height of the inner cylinder immersed in the liquid.

But many terms in the above equation may be reduced to a single apparatus constant K, so we may write

$$\eta = KM/\Omega$$ (Equn. II-5a)

Of course even in this apparatus, the rate of shear is not constant across the gap; but, if the gap is small, the variations may often be neglected.

For still higher accuracy, the cylinders may be replaced by a plate on

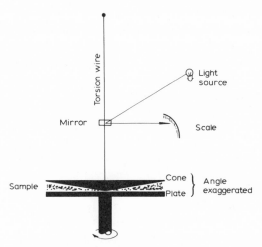

Fig. II (2). Cone-plate rheometer.

which the sample rests, and a suspended wide-angled cone. The tip is usually cut off and the bottom of the truncated cone almost touches the plate. This is shown diagrammatically in Fig. II-2. This set-up has the advantage that very small samples can be used. (The model best known to biorheologists is the "Wells–Brookfield cone-plate viscometer" [1].) One might suppose that the shear rate would be really constant whatever the angle of the cone; since if one moves twice as far from the centre, the width of the gap becomes twice as great. However, for rather complex hydrodynamic reasons, this is not the case and very flat cones must be used. If the liquid is liable to run off the plate, this may be replaced by a second cone only slightly wider in angle than the upper one [2].

It happens that there is a striking mathematical similarity between the equations which best describe the behaviour of visco-elastic systems and those used for complex alternating electric currents.

A very ingenious application of this is now in common use by rheologists.

Modifying one of the methods just described, the lower cylinder, cone or plate is caused to oscillate with a regular wave motion (a "sine curve") generally by means of a cam. The upper component is attached to a long metal wire or thread on which is fastened a small mirror as before. If a completely (soft) elastic material filled the gap, the torque would be transmitted to the upper component so as to twist the wire and a beam of light reflected from the mirror could be made to register the torque. (In fact, of course, the instruments actually used are rather more sophisticated.)

The curve registering the torque would also have the form of a wave of the same wavelength as that applied to the lower component.

At the middle of the "swing" there is no stress on the upper component: at the reversal points the stress is maximum. Thus, for an elastic material, the two wave curves will be "in phase", i.e. the maxima and minima will occur at the same times. (We call this "phase angle zero".)

For a highly viscous material, although the strain (deformation) is zero at the mid-point of the swing, it is here that the rate of flow is greatest, this being zero at the end of the swings. Elasticity (or more correctly, the shear modulus of elasticity*) is defined as the ratio of stress to strain; and the viscosity as that of stress to rate of change of strain. The rate of change of strain, and hence the stress, will be greatest at the mid-point of the swing. So, for a viscous system, the two wave curves will be precisely out of phase. (We call this "phase angle 90°"). For a linear visco-elastic system the stress curve will also be sinusoidal and the phase angle will lie between 0° and 90°. For a system represented by more than one viscous and one elastic element, so long as we keep to these simple "linear" models, the constants for the individual elements should be calculable from the variations in the complex modulus with changes of frequency.

Following the principles of Descartes, it is customary to describe data on graphs to the right of the origin as "positive" and to the left as "negative"; but, as the present writer has shown (Scott Blair [3]), this is sometimes misleading and, for simple systems, unnecessary.

The biorheologist will often come across, in the literature, the equation

$$G^\star = G' + iG'' \qquad \text{(Equn. II-6)}$$

$G^\star$ is what is called "the complex modulus" and includes the ordinary elastic modulus G' and the "imaginary part of the modulus", iG'', called "imaginary" because "i" is the square root of minus one and is an "imaginary number". G'' is really a measure of viscosity; and in simpler cases at any rate, this equation need not be used. We do, in fact, ignore the sign, treating $G^\star$ as a "positive vector", i.e. having direction (the phase angle) as well as magnitude. To indicate this, we write $|\,G^\star\,|$. This is the ratio of the maximum stress to the maximum strain.

Now we make a graph, as shown in Fig. II-3 in which G' marks the horizontal axis and G'' marks the vertical axis. To convert G'' to a viscosity ν, the frequency must be introduced because the *rate* of strain, from which

---

* The term "elasticity" can be confusing to the layman. Thus rubber is much more "elastic" (i.e. "springy") than is concrete; but concrete has a much higher elastic modulus than has rubber.

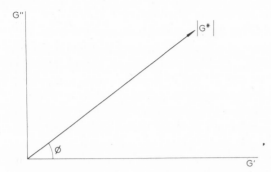

Fig. II (3). Vector diagram for complex modulus.

viscosity is calculated, depends on it and also because G and viscosity have different dimensions. A straight line is drawn, from the origin of length $| G^\star |$, making an angle ϕ with the horizontal equal to the phase angle.

Now it follows that when $\phi = 0°$, $| G^\star |$ becomes the ordinary elastic part of the "complex modulus"; and that when $\phi = 90°$, it becomes G'', which, divided by $z \pi v$ gives the viscosity (η'). For intermediate values we have, very simply:

$$G'' = | G^\star | \sin \phi \qquad\qquad \text{(Equn. II-7)}$$

and

$$G' = | G^\star | \cos \phi \qquad\qquad \text{(Equn. II-8)}$$

If v is the frequency (in cycles per second), $\omega = 2 \pi v$ is the "angular velocity". The "linear elements" may be represented by combinations of dashpots and springs, but their respective viscosities and elastic moduli will not be independent of frequency. It is only by studying the variations in moduli at different frequencies that one can tell whether the dashpots and springs should be in series or in parallel (see the book by Van Wazer et al. listed at the end of this chapter).

The data may equally well be expressed in terms of complex, real and "imaginary" viscosities. The real viscosity is got by multiplying G'' by ω.

These equations enable us to evaluate the viscous and elastic parts of the complex modulus separately.

If the model is more complex, i.e. cannot be represented by simple linear elements, the curve of the torque will not be a simple sine wave. It can be reduced to a series of sine waves by means of a rather complicated mathematical process known as a "Fourier analysis".

A generalized form of the rheometers just described was designed by Weissenberg [4]. It can be used either for rotation or for sinusoidal oscillation. This claims to measure stresses "at all angles" and is therefore called the "rheogoniometer". When certain complex materials are sheared between two cylinders (in rotation), although all the forces are applied in the horizontal plane, a force at right angles is produced (called the "normal" force) which causes the material to climb up the inner cylinder if it is not totally immersed. This phenomenon is known as the "Weissenberg effect". The rheogoniometer, which can operate with either coaxial cylinders or cone-plates, also measures the "upthrust", so that stresses can be determined "at all angles". It is undoubtedly one of the best "all purpose" type of instrument for rheological testing and is now on the market in various forms.

There are other types of rheometers that measure complex moduli. In one of these (see Birnboim and Ferry [5], and Fukada and Date [6]) a vertical rod is oscillated sinusoidally within a hollow cylinder containing the material. In another (invented by A. Képès but apparently not described in any scientific journal, though available commercially*), the sample is placed between a spherical bob and a concentric hemispherical cup, which rotates in the same direction at the same speed. The driving shaft of the cup can be swivelled through a small angle. The plane on which the two shafts lie can be turned about the horizontal axis. The phase angle is given by the angle in this plane (Fig. II-4). Other workers have used a rather simpler apparatus which consists of eccentric parallel rotating disks; but not, so far as is known, yet used for biological systems.

In general rheology, the third best-known method of testing materials is probably that of a falling (or rising, rolling or sliding) sphere. In the simplest case, a metal sphere, such as a ball bearing, or a sphere of plastic, is timed as it falls (or sometimes rises) through a known distance in a liquid. (If the liquid is opaque, the passage of the ball can be indicated electromagnetically.) G. G. Stokes gave his name to the general equation of fall:

$$v = \frac{2\,r^2\,\Delta\rho\,g}{9\,\eta} \qquad \text{(Equn. II-8)}$$

where v is the velocity of fall, r is the radius of the sphere, $\Delta\rho$ is the difference between the densities of the sphere and of the liquid and g is, as usual, the gravitational constant. Unfortunately, very large corrections have to be made in using Stokes' equation unless the size of the container, and hence the quantity of sample, is very large compared with the size of the sphere.

---

* From Contraves, Schaffhausenstrasse 580, C. H. 8052, Zürich, Switzerland.

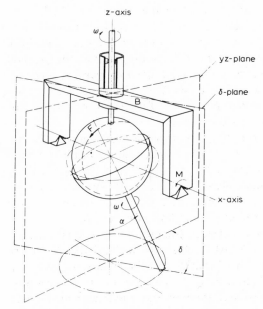

Fig. II (4). Balance-rheometer (Contraves A.G., Zürich).

More complex equations have been proposed, or empirical data have been obtained under conditions where the sphere has almost as large a radius as the container. Quite often the whole apparatus is tilted so that the sphere rolls or slides down an inclined surface. But with techniques quite large corrections must often be made for inertia effects around the sphere.

This does not, at first sight, seem to be a very suitable method for the very small samples generally available in biorheology, but a few ingenious modifications have been made which could well be applied. Behar and Frei [7] have described such an apparatus in which a tiny steel sphere falls through the material to be tested. It can be dragged sidewards by a magnet. On release, the elastic recovery can be measured. The position of the sphere is observed by means of microphotography, using "pulsed" light. A somewhat similar apparatus, but requiring a larger sample, was described by Hart [8].

In Chapter III, experiments are described in which the rising or sinking of small, naturally occurring spherical particles are measured. Stokes' equation assumes that the particles are well separated. When large numbers of particles are close together, a correction was long ago proposed by Cunningham [9]. Equn. II-8 is modified, dividing the right-hand side by a complex expression involving *r* and half the distance between the spheres. As we shall see in Chapter III, this correction has been used in work on protoplasm.

We shall see in later chapters that a number of other methods have been used by biorheologists: sinking of perforated disks, extension or compression of cylindrical test pieces, compression or shear between parallel plates or compression of the sample in bulk. But these are special cases and need not be discussed here. The electromagnetic method for measuring the rate of flow of blood in vivo will be described in Chapter V.

"Spinability", still often known by its original German name as "Spinn-barkeit", would seem, at first sight, to be easily measured by drawing out a thread of the material, either horizontally on a frictionless surface (mercury or a set of very carefully made rollers), or vertically by pulling a thread upwards attached to two grips. It is generally found that the length of the thread at breaking-point is proportional to the rate of extension over a reasonable range. We shall see, however, in Chapter VII, that it is often not easy to "grip" spinable materials. Also, it is now realized that "mercury baths" must be treated with caution. The author used one for some years in a well-ventilated laboratory, with no observable ill effects, but there have been quite serious cases of poisoning from continuous exposure to mercury vapour.

A number of other empirical test methods will be mentioned in the following chapters but need not be discussed here. For a general review article on biorheological testing, see Scott Blair [10].

REFERENCES

[1] Wells, R. E., Denton, R. and Merrill, E. W., *J. lab. clin. Med.*, 57: 646 (1961).
[2] Dintenfass, L., *Biorheology*, 1: 91 (1963).
[3] Scott Blair, G. W., *Rheol. Acta*, 11: 238 (1972).
[4] Weissenberg, K., *Proc. 1st int. Congr. Rheol.*, Scheveningen, 1948, Part II, p. 114 (North-Holland Publ. Co., Amsterdam, 1949).
[5] Birnboim, M. H. and Ferry, J. D., *J. appl. Phys.*, 32: 2305 (1961).
[6] Fukada, E. and Date, M., *Biorheology*, 1: 101 (1963).
[7] Behar, Y. and Frei, E. H., *Bull. Res. Counc. Israel*, 5A (1): 82 (1955).
[8] Hart, J., *J. sci. Instrum.*, 31: 182 (1954).
[9] Cunningham, E., *Proc. roy. Soc. A*, 83: 357 (1910).
[10] Scott Blair, G. W., *Lab. Pract.*, 15: 48 (1966).

BOOKS FOR FURTHER READING

Merrington, A. C., *Viscometry* (Arnold, London, 1949).
Scott Blair, G. W., *Elementary Rheology* (Academic Press, London, 1969).
Van Wazer, J. R., Lyons, J. W., Kim, K. Y. and Colwell, R. E., *Viscosity and Flow Measurement* (Interscience, New York, 1963).

Chapter III

RHEOLOGY OF PROTOPLASM

It is fitting that protoplasm should form the subject of the first chapter of this book that deals with specific biorheological systems, both because protoplasm comprises the most general of all biological systems and also because, in its various forms, it shows almost all the rheological modes of behaviour that have been found in other fields.

The literature runs into many hundreds of references and, as elsewhere in this book, no attempt will be made to give a complete list. In the "key papers" quoted, the reader should find most of the further references that he may need.

The best short introduction to the subject is probably to be found in a General Lecture given by Copley [1] to the 1st International Congress on Rheology. Copley defines the word "protoplasm" broadly as "the sum of materials which make up the life unit of the cell".

Perhaps one should amplify a little: all the *contents* of living cells, not including the cell membrane which surrounds them (the cortex) are included. The term "protoplasm" was probably first introduced by Purkinje in 1840 but observations were made by Dujardin in 1835. Mould, in 1846, defined protoplasm as "the contents of the plant cell", but the term is not, of course, confined to plants. From early times, much work has been published "on the viscosity of protoplasm", the viscosity values given ranging from a few centipoise to infinity!

It was the late Prof. L. V. Heilbrunn who wrote that "one might as well make viscosity measurements of a solution and include in the value obtained the viscosity of the bottle which contained it" (though the analogy is not quite exact). Those new to this field are warned against possible confusion resulting from the very similar names of two of the most prolific workers in this field: L. V. Heilbrunn, just quoted, and A. L. Heilbronn.

To return to Copley's summary, he quotes many of the most important workers up to 1949. Of special significance are Seifriz [2] and Kamiya [3].

In later years, an outstanding worker was Pfeiffer [4]*.

The article by Kamiya is also to be found in "Plasmatologia VIII" also published by Springer-Verlag and it gives a splendid review of the whole subject. Strangely similar is the review in the same series by Heilbrunn [5]. Much of the information in this chapter is taken from this latter excellent Monograph which is not easily available. But one should warn readers who go to the original that the author, though a leading authority on protoplasm, was not a rheologist. Also, for certain unfortunate reasons,. he entirely ignored the pioneer work of Seifriz, in spite of giving some 300 references to other work. (Seifriz's review article [2] was remarkably good, but is, of course, now somewhat out of date.)

The contents of a cell consist of a continuous liquid phase, the cytoplasm, and various types of particles, granules, membraneous structures etc., suspended in it**.

Protoplasm behaves in some ways very similarly to blood: it tends to coagulate when removed from the cell. It is advantageous, therefore, to measure its rheological properties (often loosely referred to as "viscosity") within the cell itself. There are two principal ways of doing this: one can measure the rate of movement either of granules existing within the cell, or one can introduce particles artificially. Surprisingly enough, this latter process does not always upset the rheological behaviour. Gravitation is seldom a sufficiently strong force to produce a convenient rate of fall and a centrifuge is generally used. The hand centrifuge has the advantage over the electrical centrifuge because, with the former, it is easier to reduce the proportion of the rather short centrifuging time that is taken up in acceleration and subsequent deceleration. Careful centrifuging does not seriously upset suitably selected sources of protoplasm. It is just as well for the astronauts that large variations in the force of gravity do not seriously affect protoplasms!

The viscosity is calculated from the force and the rate of movement of the particles. Usually, the only "correction" to Stokes' equation that is made is that mentioned in Chapter II as proposed by Cunningham to allow for the fact that in protoplasm the small spherical granules are present in large quantities.

The specific gravity of the granules, which comes into the equation, is determined by floating them in sugar solutions of different concentrations and finding the specific gravity of the solution in which they neither rise nor fall. The granules may be either heavier or lighter than the surrounding

---

* Only the most important contributions of these authors are given in this chapter.
** I am indebted to my physiologist friend, Dr. G. Patrick, for the correct terms for these components of cells.

medium and may therefore either fall or rise under centrifugal forces. This method, though generally applicable to animal cells, can seldom be used with plant cells, since these tend to clump together and to fall (or rise) in a solid mass, in which case the movement of starch grains (present in some plant protoplasm) or other inclusions, must be measured. The rigidity of the cortex must also be allowed for in certain cases, since the granules may be initially anchored very firmly to it.

The second principal method for measuring the "viscosity" of protoplasm depends on an equation originally proposed by Einstein (as a side-line in the same year as he published his Special Theory of Relativity: 1905!). This was later improved by von Smoluchowski. The original equation is:

$$D_x^2 = 14.7 \times 10^{-18}. \frac{T t}{\eta a} \qquad \text{(Equn. III-1)}$$

when D_x is the displacement of a particle in time t, a is the diameter of the particle and T the absolute temperature. In the von Smoluchowski modification, the value of D_x^2 ismultiplied by about 1.2. Even so, there are many simplifying assumptions and the values of viscosity obtained can be only approximate. However, since the values for the viscosity of cytoplasm differ so widely in the literature, even an approximate value has some significance. The usual technique is to centrifuge the granules to one side of the cell and measure their return.

Some workers have used a different formula, as proposed by Fürth [6]:

$$\eta = 1.4 \times 10^{-16}. \frac{T t}{3 \pi a l^2 n} \qquad \text{(Equn. III-2)}$$

where l is the shift in the position of the particle, and n is the number of times this distance is traversed in time t^*. But this method has seldom been used in recent times. (For a good account of this and other methods see Harvey [7].)

Using plant protoplasm Heilbronn [8] has followed in movement of nickel particles (which do not seem to disturb the protoplasm greatly) in a magnetic field.

The same worker also inserted small iron rods into large masses of slime mould protoplasm and measured the current needed to twist the rods, comparing this with the current needed for water; but the method was not altogether successful since the measured current was not independent of the size of the drops.

Studies of the process of lysis (rupture) of the granules also give at least

---

* For physical chemists: the initial constant given is an approximate value for R/N.

a rough idea of the consistency of the cytoplasm. It is often difficult to distinguish the rheological properties of the contents of the cell from those of the cell membrane, nor is it always clear whether data refer to the whole of the cell contents or to the cytoplasm only. With all these methods, it must be remembered that cutting sections for microscopic examination will often itself change the properties of the protoplasm. Such changes, not known to earlier workers, may well account for some of the wildly different values given for viscosity. Also, in some cases many "corrections" would be needed if Stokes' law is applied; not only that of Cunningham.

Measurements by Pfeiffer [9] of flow in artificial capillary tubes are also open to the criticism that the consistency may well have changed considerably.

As we shall see later, protoplasm, in its different forms, shows a wide variety of complex rheological behaviour. In so far as it may be regarded as a Newtonian liquid (see Chapter I), Heilbrunn [5], in 1958, concluded that, in plant cells, the entire protoplasm, including granules, the viscosity may be several times as great as the approximate 5 cp (i.e. five times the viscosity of water) which he proposes for the granule-free cytoplasm (see below).

Protozoan protoplasm has been studied mainly from two sources: from Amoeba and from Paramecium. A giant amoeba, which goes by the extraordinary name of *Chaos chaos*, has been much used. This has an extremely thick cortex, and for this reason, measurements of the consistency of the contents are subject to criticism (Heilbrunn [5]). This author preferred to use the much thinner-skinned *Amoeba dubia*. His results agreed closely with those of an earlier worker (Pekarek [10]). But, in work with other amoebae, very wide differences are evident between the results from different workers.

The protoplasm of Paramecium is much more complex. Heilbrunn [5] describes dense suspensions of granules, resulting in various viscous anomalies. The early published values for "viscosity" vary very widely. Although partly due to errors in calculation by one or two authors, it is also definite that the "viscosity" varies widely with the rate of shear. Larson (quoted from Heilbrunn [5] who does not give a reference) found a most unusual proportionality between viscosity and applied stress for iron, starch and carmine particles, a phenomenon which Heilbrunn [5] ascribes to "dilatancy". In Chapter I, we mentioned systems which increase their consistency when sheared. If recovery is appreciably slow, this is properly called "negative or inverse thixotropy". If recovery is too rapid to be observed under the experimental conditions, the proper term is "shear thickening". If a system of closely packed particles suspended in a liquid is sheared, the particles will move into open packing before they can pass one another. This is rightly called "dilatancy": but the term is often loosely used simply to imply shear thickening which may have quite different causes (See Chapter I). For example, Jordan [11], working with *muscle protoplasm*, describes what he calls the "snow-plough effect" in which masses of particles tend to build up in front of any object moving through them. The same phenomenon was widely studied for other systems by Röder [12] and it seems likely that it is this rather than true dilatancy that accounts for a part at least of the strange behaviour of Paramecium protoplasm. Some samples show a "viscosity" which at first increases with increasing gravitational force, and then, passing through a maximum, diminishes. It is certain that protoplasm is also often thixotropic (see ref. 7).

It is sometimes claimed that Kühne's observations of a nematode worm within the protoplasm of frog muscle fibre in 1863 marked the true discovery of thixotropy. It is uncertain whether he was really the first to observe this very common phenomenon. The rheology of muscular contraction will be discussed in a later chapter: here we shall consider only the properties of the protoplasm as such. Unfortunately, muscle protoplasm is extremely sensitive to injury. Its behaviour in producing fibrils is somewhat reminiscent of the production of pseudopodia by platelets in blood when taken from the body. (We shall see later that the behaviour of protoplasm is in many ways similar to that of blood, especially in coagulation.)

Some fairly successful experiments were made by Rieser [13] who injected small oil drops into the muscle fibres of the frog and measured the speed with which the droplets rose. But even when this could be done without much apparent damage to the protoplasm the values of viscosity obtained varied from 14 to 280 cp, and these variations showed no clear relationship to the nature of the oil used. Others have tried, with even less success, to repeat Rieser's experiments.

Heilbrunn [5] concludes "that a striated muscle fibre consists of an outer stiff cortex and an inner mass of protoplasm which is fluid. In all probability, this inner protoplasm is even more fluid than Rieser's measurements indicate, for even in his best experiments there must have been some injury, with subsequent increase in viscosity." *Nerve protoplasm* is even more difficult to study, because of clotting.

A very strange type of organism, much investigated, comprises the *slime moulds**. An excellent review article on their rheology is written by Jahn and

---

* Mycetozoa, or Myxomycetes: these are a very unusual type of organism. They form into groups in a column, initially standing vertically, then falling over to lie in a sausage-like form. This whole "sausage" migrates across the forest floor to some place where there is more food. It then disintegrates into, perhaps, half a million individual amoeboid cells who proceed to feed on the local bacteria until stocks run short, when the coalescence is resumed, and a further migration occurs.

26

his colleagues [14, 15]. Since these papers include nearly a hundred references, only a brief summary can be given here.

These strange organisms when "integrated" consist of a sponge-like structure (plasmodium) which is neither sol nor gel, in the sense that cross-linking bonds are present in quite variable proportions. (In a sense, this is not unlike the situation in the early stages of coagulation of blood or indeed of protoplasm.) There are also long tubes the walls of which expand and contract, causing rapid changes in both the rate and direction of flow of the liquid they contain. Jahn [15] has studied the hydrodynamics of this process and discusses theories offered by many earlier authors. Some of the earliest work was done by Seifriz [2] but, in more recent times, "key" papers which readers may care to study, are those of Stewart and Stewart [16].

As one would expect, these materials show many rheological anomalies: shear thickening, shear thinning, thixotropy, etc. Jahn's papers show diagrammatic pictures of the distension and contraction of capillaries and cells. More information is given in the monograph by Kamiya [3] already cited: indeed much recent work has been done in Japan. At the 4th International Congress on Rheology, interesting papers were given by Abe [17] and by Kamiya and Kuroda [18], also a short note by Allen [19].

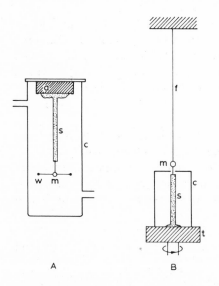

Fig. III (1). Abe's apparatus for measuring the viscoelasticity of slime mould protoplasm. Reproduced from: Proceedings of the 4th International Congress on Rheology, 1963, Part IV (Ed. A.L. Copley), p. 147, Interscience, New York and London (1965).

In the first of these papers, Abe prepared smooth strands of plasmodium (from *Physarum polycephalum*). These were from about 6 to 10 mm long and had diameters of about 500–750 μ. Working with Seifriz in 1954, Kamiya had already shown that these strange strands, if hung vertically in a moist atmosphere, will oscillate with a constant period without the application of any external torque.

But in these later experiments, Abe designed two very neat instruments to measure the damping of oscillations following an applied torque. These are shown in Fig. III-1.

In the first (marked A), the strand (s) is drawn from a mass of agar gel (a), to which it adheres, and is attached at the bottom end to a kind of "micro-pendulum" (m) bearing a mirror which reflects a beam of light to register the oscillations. The whole apparatus is placed in a moist containing vessel. In the second apparatus (B), (f) is a glass fibre, with mirror (m) attached. The lower end of the sample adheres to a turntable (t) whose angle of twist can be measured.

A chart from the first apparatus, showing the damped oscillation, is given in Fig. III-2.

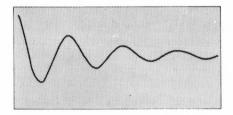

Fig. III (2). Chart from apparatus shown in Fig. III-1 showing damped oscillations. Reproduced from: Proceedings of the 4th International Congress on Rheology, 1963, Part IV (Ed. A.L. Copley), p. 147, Interscience, New York and London (1965).

From the data obtained by these methods, Abe concludes that the strands follow what the rheologist calls a "Burgers model": i.e. a dashpot and spring in parallel attached in series to a second dashpot and spring in series.

The elastic moduli of the "springs" and the viscosities of the "dashpots" can be calculated. In two experiments using Method B, the series pairs gave moduli of 1×10^4 and 5×10^4 dynes/cm^2 and viscosities of 5.2×10^6 and 12.0×10^6 poise: the "parallel" pair gave values of 2.5×10^4 and 6.6×10^4 and $2 \ (2.0?) \times 10^6$ and 2.2×10^6, respectively.

Repeated experiments showed that the strands were thixotropic. These values are extremely high, especially the viscosities. The author points out, however, that the plasmodial strand which he used, consists far more of ectoplasm (cortex) than of the rest of the cytoplasm (endoplasm). Method A gives only one elastic modulus (ave. 2.8×10^4 dynes/cm²) and one viscosity (ave. 4×10^3 poise), the latter unexplainably lower than the values obtained by Method B, though the author hints that this may be due to thixotropy.

This author worked in cooperation with Prof. Kamiya, an author of the second paper, which is concerned with the velocity profile of endoplasmic streaming of a rhizoid cell of *Nitella flexilis*. For lack of material, it was not possible to study flow through an artificial capillary of protoplasm as a whole, but for isolated endoplasm, this could be done. A special viscometer was designed for this purpose, using an agar capillary.

Profile curves of a typical "flattened" paraboloid type were obtained. There is considerable slip between the agar wall and the endoplasm surface. The shear rate vs. stress diagram shows both curvature and a yield value. The authors conclude that "the motive force responsible for the streaming [under its own power] is an active shearing force, or parallel shifting force, generated at the boundary between the cortical cytoplasmic gel layer and the outer edge of the endoplasm".

In the short third paper, Allen studied the movement of iron spheres in an acellular slime mould. There were marked and complex changes in velocity as the spheres moved through the sample, indicated a complex inhomogeneity.

We leave till last in this section what is perhaps the commonest source of protoplasm for rheological studies: *marine eggs*, e.g. sea urchins, starfish, worms and clams. Heilbrunn [5] writes: "The egg of *Arbacia punculata* (Lamarck) has become a standard object of study." It will not be possible here to give more than a very brief survey of this work. For further information, the reader is referred to a book by Ethel B. Harvey [20]. Heilbrunn [21] himself, as early as 1926, made the first quantitative measurements in any animal cell using this egg. The speed of motion of these cells can be fairly easily measured using the centrifuge, but they elongate as they move and a suitable correction for this must be made in applying Stokes' equation; as well as Cunningham's correction. The viscosity was found to be 1.8 cp using the colourless granules that constitute a majority of the inclusions. These granules of various sizes interfere with one another and the value for the viscosity of the entire protoplasm is given as 6.9 cp. But these values can be only approximate and there is poor agreement between the values found by different workers.

Using the alternative Brownian movement method, Heilbrunn got values for viscosity of about 5 cp. Such researches are complicated by the fact that the consistency of the protoplasm varies widely, even for eggs from the same species, being affected even by the degree of contamination of the water from which they are taken. The experimental variations are large compared with any differences in viscosity due to differences in shear rate, since it has been shown that, for these eggs, the protoplasm is nearly Newtonian (though in some cases, thixotropy has been reported). There are, however, rather dramatic changes in viscosity during the first minute or so after the egg leaves the ovary and enters the water.

Much work was published by Pfeiffer [22, 23] (only two of the key papers are listed here) on the opticorheological properties and also the "Spinnbarkeit" ("Spinability": capacity to be drawn out into long threads) of protoplasm and similar materials. *Rheodichroism* is defined as dichroism induced by flow, where "dichroism" means "that phenomenon in which the absorption by a transmitting material of light of a given wavelength varies with the state of polarization of the transmitted light" (ref. 23, p. 146). Dichroism is said to be "positive" when it shows maximum absorption of polarized light in a plane parallel to a given axis, and "negative" when it is perpendicular. The polarization may be either linear or elliptical.

The structural units, which are too small to be seen with a microscope with visible light, Pfeiffer calls "leptones". His first studies on rheodichroism were applied to blood plasma, but later, he studied protoplasm. Polarized light absorption along and across the direction of flow of coloured leptones in a capillary was measured. The "Spinnbarkeit" was also measured. The rheodichroism was related to stress–shear rate data and these, in turn, to the "Barus effect" (swelling of an extruded cylinder). Data from rheodichroism are very similar to those derived from measurements of flow birefringence.

Ageing of cells. Heilbrunn [5] quotes a review article by Fischer [24]. Again, there is much confusion in the literature, some authors claiming a rise and others a fall in viscosity with ageing (materials from plant cells). Again comes the problem of what parts of a heterogeneous system are included in the sample studied. It seems likely that the interior protoplasm increases in viscosity but that the cortex (or ectoplasm) loses consistency. Protoplasm from sea urchin eggs seems to show an overall increase in viscosity. With aquatic plants, the viscosity varies with the season, and even the time of day. For the effects of variations in temperature of salts, acids and bases and of radiation, the reader should consult Heilbrunn's review [5].

Clotting of protoplasm. Although less is known about the complex processes of protoplasm clotting than about the clotting of blood (see Chapter V), these processes must have much in common. In both cases, a series of, mainly enzyme, reactions takes place and in both cases, calcium plays a major role. Likewise the usual blood anticoagulants, notably heparin, also prevent coagulation of protoplasmic material. Vitamin K and dicoumarin behave much as they do in respect of blood, but also greatly affect cell division. The effects of the former on changes in viscosity following fertilization follow a different pattern from that for untreated protoplasm, the viscosity passing through a maximum in the course of an hour or two. Vitamin K acts in the opposite fashion.

We close this chapter with a brief account of the effects of anticarcinogenic substances on protoplasm (this is one of the, unfortunately, very few fields in which rheology may be of use in the study of cancer). Anticarcinogenic substances, like effective radiations, generally have the property that they prevent or slow down the growth of cancer cells in small doses but may produce neoplasms when used in larger doses. Nitrogen mustard hydrochloride prevents cell division by keeping the protoplasm fluid. Nitromin has a similar effect when used in small quantities but exactly the opposite effect in large doses: urethane behaves somewhat similarly. These reactions are all connected in some complex way with clotting. Even heparin has some effect in improving the survival time of cancerous animals. A bacterial polysaccharide from *Serratia marcescens* is a less powerful anticoagulant and has been used for cancer therapy. Many other heparin-like substances have been tried. It is to be hoped that further studies of their effects on protoplasmic consistency will prove of value. (A full account of work up until 1958 is given in Heilbrunn's article [5].) A major difficulty in this field is that many substances which diminish the consistency of the interior of the protoplasm, increase that of the exterior—and vice versa.

REFERENCES

[1] Copley, A. L., *Proc. 1st int. Congr. Rheol.*, Scheveningen, 1948, Part I, p. 47 (North-Holland Publ. Co., Amsterdam, 1949).
[2] Seifriz, W., in *Deformation and Flow in Biological Systems* (Ed. A. Frey-Wyssling), p. 3 (North-Holland Publ. Co., Amsterdam, 1952).
[3] Kamiya, N., *Protoplasmic Streaming* (Springer-Verlag, Vienna, 1959).
[4] Pfeiffer, H. H., Paper in *Flow Properties of Blood and Other Biological Systems* (Eds. A. L. Copley and G. Stainsby) (Pergamon Press, Oxford, 1960).
[5] Heilbrunn, L. V., *Plasmatologia*, 2 (1) (1958) (Monograph: *The Viscosity of Protoplasm*).
[6] Fürth, R., *Z. Phys.*, 60: 313 (1930).
[7] Harvey, E. N., *J. appl. Phys.*, 9: 68 (1938).
[8] Heilbronn, A. L., *Jb. wiss. Bot.*, 61: 284 (1922).

[9] Pfeiffer, H. H., *Protoplasma*, 33: 311 (1939) and 34: 347 (1940).

[10] Pekarek, J., many papers, but see especially *Protoplasma*, 11: 19 (1930) and 17:1 (1932).

[11] Jordan, J., *Jb. Morphol. mikrosk. Anat.*, 45 (Part 2): 46 (1939).

[12] Röder, H. L., *Rheology of Suspensions* (H. J. Paris, Amsterdam, 1939).

[13] Rieser, P., *Protoplasma*, 39: 95 (1949).

[14] Jahn, T. K., Rinaldi, R. A. and Brown, M., *Biorheology*, 2: 123 (1964).

[15] Jahn, T. K., *Biorheology*, 2: 133 (1964).

[16] Stewart, P. A. and Stewart, B. T., *Exp. Cell Res.*, 17: 44 (1959) and 18: 44 (1959).

[17] Abe, S., in *Symposium on Biorheology, Proc. 4th int. Congr. Rheol.*, Providence, R. I., 1963, Part IV (Ed. A. Copley), p. 147 (Interscience, New York, 1965).

[18] Kamiya, N. and Kuroda, K., in *Symposium on Biorheology, Proc. 4th. int. Congr. Rheol.*, Providence, R. I., 1963, Part IV (Ed. A. Copley), p. 157 (Interscience, New York, 1965).

[19] Allen, R. D., in *Symposium on Biorheology, Proc. 4th int. Congr. Rheol.*, Providence, R. I., 1963, Part IV (Ed. A. Copley), p. 173 (Interscience, New York, 1965).

[20] Harvey, E. B., *The American Arbacia and Other Sea Urchins* (Princeton Univ. Press, Princeton, N. J., 1956).

[21] Heilbrunn, L. V., *J. exp. Zool.*, 43: 313 (1926).

[22] Pfeiffer, H. H., *Kolloid Z.*, 136: 156 (1954) and 147: 53 (1956).

[23] Pfeiffer, H. H., Paper in *Flow Properties of Blood and Other Biological Systems*, (Eds. A. L. Copley and G. Stainsby) (Pergamon Press, Oxford, 1960).

[24] Fischer, H., *Protoplasma*, 39: 661 (1950).

Chapter IV

RHEOLOGY OF MUSCLE AND ITS PROTEINS; COLLAGEN; BONES; BRAIN INJURIES; HAIR; MECHANOCHEMISTRY

Muscle

So much work has been published on the rheology of muscle, myosin and actin that it will not be possible to give a complete bibliography, even of the important papers, in an introductory book of this kind. Moreover, most of the papers which include some rheology, are mainly biochemical. An attempt will be made, however, to provide adequate sources for further study. (For a review of some very recent papers, see the Appendix.)

More than half of the protein substance of muscle consists of myosin. This can be extracted by salt solutions such as KCl, and it may exist either in an amorphous or crystalline form, or as a tactoid*. After further extraction, the myosin combines with the second protein, actin, to form actomyosin. Myosin catalyses the hydrolysis of adenosine triphosphate: the famous "ATP". Glycerol-extracted muscle fibres, in which the contraction in many ways resembles that of living muscle, can be prepared. (This brief account is condensed from an interesting review article by Mommaerts [2]. See also another article by the same author [3] and papers by Szent-Györgyi [4, 5].)

The properties of myosin and actomyosin have formed the subject of too many authors to quote in extenso, but see especially von Muralt and Edsall [6]; Symposium of the New York Academy of Sciences in 1947 (Vol. 47 of Annals); Meyer and Picken [7] and a series of papers in Proc. roy. soc. B, Vols. 151, 152 and 155 (1959–1962) in which the visco-elastic properties of muscle are discussed; Greenstein and Edsall [8]; and Dainty et al. [9]. These last authors showed that, when myosin** solution is treated with ATP, its

---

* Tactoid: "a material that shows patches of oriented colloidal particles, usually eye-shaped, within a liquid", see Reiner and Scott Blair [1].

** *Postscript*: Dr. Dorothy Needham has pointed out (private communication) that this early work was done before the discovery of actin. An up-to-date modification of the conclusions is given in her recent book, "Machina Carnis", pp. 146–150 (Cambridge Univ. Press, Cambridge, 1972).

flow birefringence is decreased, its anomalous viscosity is retained and its relative viscosity is decreased. A brief historical survey shows Wöhlisch [10] as one of the first to compare the contraction of collagen (to be discussed below), which forms an important part of animal tissue, with that of rubber and similar discussions have taken place concerning muscle. In 1927, Freundlich [11] suggested that the interior of muscle cells behaves like a thixotropic gel.

In 1933, Karrer [12] also compared the elasticity of muscle with that of rubber. Muscle cells consist of myofibrils bathed in protoplasmic liquid called "sacroplasm". Fenn and Marsh [13], in 1935, made a study of the rheology of muscle on classical lines and attempted to propose dashpot–spring models, but found this extremely difficult. The dashpots could not be Newtonian.

Meyer and Picken [7] again compared the behaviour of muscle with that of rubber. The temperature coefficient of the elastic force is reduced by stretching to a constant length but at intermediate strains, it is positive. They concluded that the structure consists of "flexible protein chains forming a three-dimensional network, and free chains in the meshes of this net". X-ray diagrams show, as with rubber, a structure for the highly strained material.

Mommaerts [14] showed that "myosin alone has a moderate viscosity not dependent on shear rate and has relatively little orientation of flow as indicated by birefringence. Actomyosin has a much higher viscosity, which is strongly dependent upon flow gradient, and birefringence, even at low shear, already indicates maximal orientation. . . . " (quoted from ref. 2. Fig. IV-1 is reproduced by permission from this paper).

Actin monomer has a molecular weight of about 60,000. On addition of salts, it becomes very viscous as a result of polymerization. There are, of course, other proteins in muscle. These have been studied by H. E. Huxley and his colleagues. In one of his later papers [15], he describes how chemical energy is converted into mechanical energy in muscle and discusses current views on the interaction between actin and myosin [15]. (This subject was being studied by A. Katchalsky whose fine work was terminated by his tragic murder by terrorists in 1972.)

The rheological properties of living muscle are extremely complex and differ for different muscles. The reader is referred to ref. 2 for further particulars. We will conclude this section with mention of a few more recent papers. (See also the Appendix.)

Paslay et al. [16] discuss "constitutive equations in theoretical muscle mechanics". Weis-Fogh and Anderson [17] studied the elasticity of elastin from bovine ligaments. This differs in behaviour fundamentally from rubber,

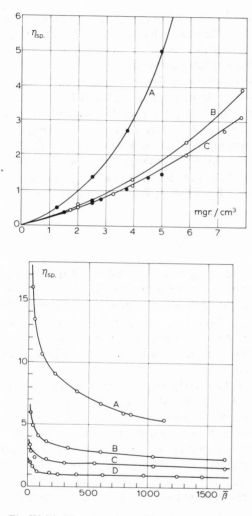

Fig. IV (1). Viscosity of myosin and actomyosin (from Mommaerts, 1945). Upper: viscosity as measured in an Ostwald capillary viscosimeter at indeterminate flow gradient, for acto-myosin (A), "myosin", actually an actomyosin of low actin content (B), and A and B in the presence of ATP. Lower: viscosity of actomyosin as a function of the velocity gradient at four different concentrations. Reproduced from: Mommaerts, W.F.H.M., Arch. Kemi Mineral Geol., 1945, nos 17, 18 and 19.

as is shown by measurements of heat exchange. Spherical globules are stretched into prolate spheroids. There are changes in both energy and entropy. Water-swollen elastin is regarded as a two-phase system. Its behaviour is essentially dependent on interfacial forces.

Blangé et al. [18] are in controversy with A. F. Huxley and R. M. Simmons concerning the role of inertia in the elastic behaviour of muscle.

Recently, a Russian worker, Deshcherevskiĭ [19], has studied the contractile properties of striated muscle. A quantitative theory is proposed depending essentially on the interaction of myosin and actin from which Hill's [20] equations follow. The theory is specifically applied to auto-oscillations, especially in insect flight muscles, and frog sartorial muscles under isotonic conditions. This is far removed from the theory of rubber elasticity.

Collagen

Collagen is one of the most interesting body materials from the point of view of the rheologist and it is somewhat surprising that relatively little work appears to have been published. Most of the published work is primarily concerned with the biochemistry. This is extremely important, but it makes for difficult reading for the rheologist who is not a biochemist.

Fortunately, there is one excellent review article by Harkness [21] which will be widely quoted in this chapter.

Although only about 5% of the total body weight of mammals consists of this protein, it is responsible for the whole stability of the body.

The chemical structure is now fairly well-known and is described by Harkness [21]. Collagen exists in the form of large numbers of individual "fibrils". These are generally grouped into bundles known as "fibres", which branch and join together to form a network. The fibrous structure enables the material to withstand high tensile stresses; indeed the strength of tendons often greatly exceeds that which can be exerted by the muscles attached to them.

In other parts of the body, where stresses are not unidirectional, a much more complex structure of collagen fibres must exist. In the case of the eye (see Maurice [22]; see also Chapter IX) the transparent cornea consists of "sheets of fibrils all running in the same direction, but the direction alters from sheet to sheet, the fibrils in each being set at right angles to those in the sheets on either side". The opaque sclera consists of bundles of fibrils running in all directions.

In the cartilage which forms the joints between the bones, about half of the material is collagen and the other half is a watery polyelectrolyte gel. Again, Harkness [21] discusses its chemical composition.

In the walls of the large arteries especially (see Chapter VI) collagen is associated with another protein, elastin. This is a material showing rubber-like elasticity and low tensile strength. It resists continuous stress better than does collagen—without "creep".

Although dashpot–spring models (or the equivalent equations) can be made for certain types of collagenous tissue, there is no general equation: power equations are also often used (see Kenedi et al. [23]).

The ultimate strength of tissues may have to allow not only for stresses produced by ordinary bodily movements but also to some extent for external shocks. Harkness quotes some remarkable differences in strength for different animals. The reasons for these differences are not obvious. Changes in size of collagenous tissue become enormous in the uterus and cervix during pregnancy and parturition. In some small animals this is even more striking than in women. Thus within the brief period of half an hour, the uterus of the rat is reduced in size from some 50 ml to negligible proportions.

The diameter of the cervix likewise increases during pregnancy so that, at parturition, it is many times its normal size, to allow for the passage of the foetus.

Harkness shows "growth curves" for such processes. These are, not surprisingly, sigmoid in shape, but the central, linear part is surprisingly long and approximate growth equations (for what these are worth) have been developed.

In some parts of the body, the reaction of the collagenous tissue must be essentially elastic: e.g. in the ejection of blood from the heart, in contrast to the "flow" characteristics which we have just discussed.

Considerable work has been done on the effects of quite large temperature changes and of various chemical substances on collagen. These are not always of immediate physiological importance but they add to our fundamental knowledge of the structure and behaviour of the material.

Not very much of importance appears to have been published since Harkness' review, except for some interesting work in the U.S.S.R. Kukhareva et al. [24] studied the effect of loading on what is called the "order–disorder transition". This work is mainly concerned with the contraction of collagen fibres when placed in water and in certain aqueous solutions and also with changes of temperature. The authors show curves which include dependence of maximal contraction on stress as well as the temperature effects. These are linear over quite a wide range.

Plotting against stress, in KCNS solution, both isometric (constant length) and KCNS concentration curves are strikingly linear.

Much of this work deals with the molecular configurations of collagen molecules and their thermodynamic aspects. There are "critical" temperatures and tensions above which a crystalline phase cannot exist. The bibliography includes references to many other Russian papers not so well-known in the West. (But see Cohen et al., APPENDIX p. 200)

Blanton and Biggs [25] have compared the tensile strength of foetal and adult human tendons. These authors are mainly concerned with the effects of embalming.

Since writing this brief account of work on the rheology of collagen, the author's attention has been drawn to a very important paper by Rigby et al. [26]. These authors measured creep and relaxation of rat tail tendons. For small extensions, over short periods, when stretched in normal saline, the tendons show simple and reproducible rheological behaviour; but over-straining produces a softening of the collagen, though it remains elastic. Extensions of over 35 % can be reached in this way. Lowering the temperature below the normal has no effect, but even quite a slight increase above 37°C causes important irreversible changes. This last finding may well be highly significant in relation to the effects of fever on the heart and other organs. "The collagen fibrils in lesions of the major collagen–vascular diseases, such as rheumatic fever, rheumatoid arthritis, etc., appear to swell, become irregular in outline and degenerate."

Bone*

Little work appears to have been done on the rheology of bone before 1942, when Bell et al. [27] measured the strength and size of rat bones in relation to calcium intake. (References are given in this paper to some less important earlier work.) Bending tests were done on femur bones, the ends of which where cast into plaster pieces. These were held horizontally, and loads were added incrementally by hanging weights from the centre of the test piece. Twisting tests were also done. There was some slow recovery when the loads were removed, but the deflexions were proportional to the load up to the point of rupture. Breaking stresses and twisting movements were also cal-culated. There was found to be a maximum calcium uptake for increasing the strength. Bone was found to be stronger than timber, nearly as strong as cast iron and half as strong as steel.

*Many papers on the anatomical study of bones are to be found in the "Journal of Bio-mechanics". These cannot be dealt with individually here.

My friend Prof. J. M. Burgers [28], quoting D'Arcy Thompson, suggests that "the elements of bone . . . arrange themselves in the direction of the lines of principal stress for the most important types of loading of the structure in which they occur . . . My opinion is that the growth of these elements of bone or of fibrous tissue at the place where they are found is not a merely physical or mechanical reaction to a state of stress: there is involved in it some form of 'recognition', some notion that is it 'worth while' to grow at such a spot and in such a direction." Nevertheless, it must not be forgotten that many materials tend to harden in the direction of applied loads.

Further experiments on the mechanical properties of bone were done by Weir et al. [29]. In this paper, the effects of rickets were quantitatively studied.

(In 1961 there was some controversy in the columns of "Nature" concerning bone rheology between Dreyer and Pearson which may be of interest to some readers [30].) The anisotropy of bone was studied by Bonfield and Li [31]. Chemical aspects were studied in a rather inaccessible paper by C. H. Lerchenthal and several colleagues at the Israel Institute of Technology, Haifa in 1968. This paper is summarized (translated from French) as follows: "Structures presenting a permanent orientation can be produced in an amorphous gel of reconstituted collagen by the application of a comparatively weak tension–force."

Somewhat earlier, Evans [32] made careful measurements of the tensile strength of bones, as part of a Symposium the rest of which will also be of interest to some readers.

Evans reminds us that, as is well-known to engineers, in a bending test, tensile stresses and strains are formed in the concave part of the specimen and compressive stresses and strains in the convex part. As early as 1876, Rauber had shown that bone is weaker in tension than in compression. Evans used a tensile strength testing technique quite similar to that familiar to engineering rheologists for many inorganic materials. The main purpose of his work seems to have been to compare the strength of bone from embalmed and unembalmed material, the latter being obtained from amputations above the knee "in which only the distal quarter of the femur was present". Wet and dry specimens were also compared. He found that the ultimate tensile strength of embalmed bone was somewhat greater than that of the unembalmed; and also that wet bone was in general weaker than dry bone. Anisotropy was also studied and compared with that of wood. Calabrisi and Smith [33] had already found that embalming *decreases* the compressive strength of bone. Other references are given in this paper to earlier work on compressive testing, but tensile tests have been few. Strengths of bones from various parts of the leg were compared.

Two quite recent papers should also be mentioned: Wall et al. [34] discuss Evans' work and compare his results with those obtained in testing such materials as cement. Many further references are given and a long list is made of the factors which influence bone strength. These factors may be grouped under six headings:

(1) particulars of the individual (age, sex, height, etc.);
(2) nature of the bone tested (location, age density, etc.);
(3) conditions of storage;
(4) preparation of sample;
(5) shape, size, etc. of test piece;
(6) method of testing.

Each of these is discussed in turn.

Finally, a paper by Wood [35] will lead us to a consideration of the rheology of head injuries. This paper is concerned with the mechanical properties of human cranial bone in tension.

The specimens were taken from compact layers of parietal temporal and frontal bone at autopsy. Strain rates covered a wide range: $0.005-150 \text{ s}^{-1}$. The elastic modulus and strength as well as the breaking strain depend on the rate of straining. The energy absorbed is not rate sensitive. There are no statistically significant differences depending on type of bone, side of body and age of subject. Unlike long bones, cranial bones appear to be transversely isotropic.

This paper is immediately followed by an article by Ommaya and Hirsch [36] on tolerances for cerebral concussion from head impact and "whiplash", a subject to which we shall now turn (Fig. IV-2).

Brain injuries

The classical paper in this field was by Holbourn [37]. His conclusions, mainly derived from "models" (made from gelatine in a paraffin wax "skull") showed concussion as essentially produced by a rotational injury. A footballer can "head" quite a heavy ball with considerable force and do no harm to himself*, whereas a "knock-out" blow in boxing may prove fatal. The brain itself is anisotropic, so that the direction of impact may be critical.

Ommaya and Hirsch, in the above-quoted paper [36], and Ommaya et al.,

---

* Though Prof. W. R. Matthews has found that such incidents can induce migraines.

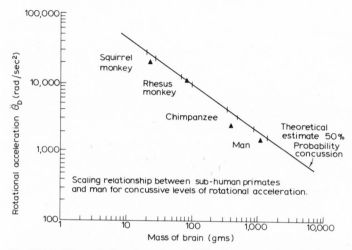

Fig. IV (2). Theoretical scaling of probability for onset of cerebral concussion in primates assuming that the crucial variable between species is mass of brain and that the crucial injury mechanism is severe shear strain imposed by brain rotation. Reproduced from: Ommaya and Hirsch, J. Biomech., 4: 20 (1971).

in earlier communications [38], only one of which is quoted here, have only partially confirmed Holbourn's findings. Ommaya and his colleagues worked on real brains, from monkeys. The effects of rotational strains of short duration were confirmed, but they found that "about twice the rotational velocity was required to produce cerebral concussion when the animal experienced indirect impact (whiplash). . . . This suggested a significant contribution to brain injury by the local effects of impact (i.e. contact phenomena)." These authors conclude that about 50% of the brain injury during impact to an unprotected moveable head is proportional to the amount of head rotation, the remaining 50% depending on skull distortion. Of course the relative sizes and strengths of brain and skull differ greatly in monkeys and in man, and even between different monkeys. Fig. IV-2 shows the mass of the brain for different monkeys and for man as related (theoretically) to the rotational acceleration.

It is hoped that further work will help in the design of protective shields for racing motorists and others especially liable to head impact accidents. A complex mathematical model for the primate brain in vivo is given by Engin and Wang [39].

Skin

Considerable work has been published on the tensile properties of skin, most of which is fairly recent. Ridge and Wright [40] stretched samples of skin before the yield value is reached and also studied stress relaxation at constant strain. The behaviour was much the same as that using highly orientated collagen from rats' tails. In the initial extension, the collagen "wicker-work" is orientated but at higher loads the collagen is itself stretched. This work confirms earlier findings that the rheology of skin varies greatly in different parts of the body. The loading and relaxation curves were plotted following empirical equations: extension vs. log load and load vs. log time, respectively. Comparisons are made with the behaviour of other body materials.

Later, Verna Wright [41] contributed an article to a Symposium on Biorheology giving further data on skin, in which the experiments described in an earlier paper were extended and further discussed.

Tests on skin in vivo were made by Finlay [42]. This author gives a good summary of earlier work for which we have no space to spare in this book. Finlay designed a complicated torsional test apparatus which made it possible to measure the frequency response of skin as a visco-elastic system (see Chapters I and II), using sinusoidal displacements. This is probably the most fundamental published study of skin rheology.

The results have some practical importance in relation to surgery. "Excessive tension necessary in sutures to close an excised area can [consequently] be reduced by applying to the sides of the wound a greater force than that required to just close the wound." This load should be maintained for several seconds. The visco-elastic properties of the skin are, however, extremely complex and the original article should be consulted. The effects of sulph-hydryl reagents on the mechanical properties of the skin are discussed by Harkness and Harkness [43].

Indian workers (Muthiah et al. [44]) have studied the skins of a number of animals and from various parts of the body: this is in connection with the preparation of leather.

In a recent paper, Sakata et al. [45] studied the mechanical properties of sheep skin under compressive stress, by means of a simple stress-strain technique and also a cyclic stress fatigue method. Skin was found to behave in a linear fashion for small loads but was non-linear for large loads. The effects of various solutions and enzymes were also studied. Fatigue behaviour is a logarithmic function of time.

Veronda and Westmann [46] performed uniaxial tests on cat skin. A comparison of force–extension curves and analytical stress–strain relations makes possible the determination of suitable strain energy functions.

Hair

The author can find little recent biorheological work on hair (as distinct from commercial wool). In 1944, Bull and Gutmann [47] studied the elasticity of keratin, the principal protein of hair, and found marked hysteresis in extension–load curves. At high extensions, there was a type of (thixotropic?) reversible softening. The authors criticize previous work by Astbury on the structure of keratin.

In the following year, Bull [48] extended this work and claimed that human hair, when extended up to 40%, showed an energy rather than an entropy type of elasticity. Later work was reported in a third paper in 1954 [49].

Meyer and Haselbach [50] made a further study of the elastic properties of hair. There was considerable controversy at about this time as to whether the elasticity of hair is of an entropy or an energy type. The difficulty lies, of course, in making sufficiently accurate measurements of very small temperature changes (see also Jagger and Speakman [51]).

Treloar [52], however, in 1951, studied the effect of tension on water absorption of hair and his results tended to bring the elastic behaviour more into line with that of rubber.

Since "biorheology" is commonly restricted to aspects of biological interest, work on wool as an industrial material will not be discussed in this book.

In a lighter vein Wright [53] studied the effects of heat and moisture on the curling of hair, with obvious reference to "permanent" waves; and, in the same year, Copley [54] made a study of baldness, measuring the force required to pull hairs out of the scalp, as well as elastic properties. He compared conditions for white Americans and Chinese and also discussed problems of the "permanent" wave.

Mechanochemistry

A word should be added about the now very important branch of study called "mechanochemistry". This work, originated by W. Kuhn, was well described in an article by Katchalsky and Oplatka [55]. The motility of animals depends on the direct conversion of chemical into mechanical energy. Many swollen macromolecular substances can make this conversion. Randomly kinked macromolecules are especially suited for this. If a reagent reacts with several groups on the chain, this is due to changes in dimensions. Electrostatic forces control the dimensions of the molecular chains; hence, changes in the degree of ionization or screening interaction of the charged groups will be transmitted along the chains, and so cause changes of shape. Such reactions suggest the application of the tools of thermodynamics but, to be effective, the thermodynamics must not be confined to closed systems. Katchalsky and others have led the way in the development of this very difficult extension of the classical theory.

REFERENCES

[1] Reiner, M., and Scott Blair, G. W., in *Rheology: Theory and Applications* (Ed. F. R. Eirich), Vol. 4, p. 461 (Academic Press, New York, 1967).

[2] Mommaerts, W. F. H. M., *Lab. Pract.*, 15: 171 (1966).

[3] Mommaerts, W. F. H. M., *Muscular Contractions* (Interscience, New York, 1950).

[4] Szent-Görgyi, A., *Stud. Inst. Med. Szeged*, 3: 76 (1943).

[5] Szent-Görgyi, A., in *The Structure and Function of Muscle* (Ed. J. H. Bourne) (Academic Press, New York, 1960).

[6] von Muralt, A. and Edsall, J. T., *J. biol. Chem.*, 89: 315 (1930) and 89: 351 (1930).

[7] Meyer, K. H. and Picken, L. E. R., *Proc. roy. Soc. B*, 124: 29 (1937).

[8] Greenstein, J. P. and Edsall, J. T., *J. biol. Chem.*, 133: 397 (1940).

[9] Dainty, M., Kleinzeller, A., Lawrence, A. S. C., Miall, M., Needham, J., Needham, D. M. and Shih-Chang-Shen, *J. gen. Physiol.*, 27: 355 (1944).

[10] Wöhlisch, E., *Verh. phys.-med. Ges. Würzb.*, 51: 53 (1926).

[11] Freundlich, H., *Protoplasma*, 2: 278 (1927).

[12] Karrer, E., *Protoplasma*, 18: 475 (1933).

[13] Fenn, W. O. and Marsh, B. S., *J. Physiol.* (Lond.), 85: 277 (1935).

[14] Mommaerts, W. F. H. M., *Arch. Kemi Mineral. Geol.*, Nos. 17, 18, 19 (1945); also *Nature*, 156: 631 (1945).

[15] Huxley, H. E., *Science*, 164: 1356 (1969).

[16] Paslay, P. R., Soechting, J. F., Stewart, P. A. and Duffy, J., *Div. biol. med. Sci.*, Brown Univ. Rep., No. 1 (1969).

[17] Weis-Fogh, T. and Anderson, S. O., *Nature*, 227: 718 (1970).

[18] Blangé, T., Kasemaker, J. M. and Kraemer, A. E. J. L., *Nature*, 237: 281 (1972).

[19] Descherevskii, V. I., *Biorheology* 7: 147 (1971).

[20] Hill, T. L., *Proc. nat. Acad. Sci.* (Wash.), 61: 889 (1968).

[21] Harkness, R. D., *Lab. Pract.*, 15: 166 (1966).

[22] Maurice, D. M., in *The Eye* (Ed. H. Davson), Vol. 1, p. 289 (Academic Press, New York, 1962).

[23] Kenedi, R. M., Gibson, J. and Daly, C. H., in *Structure and Function of Connective and Skeletal Tissue* (NATO Advanced Study Inst., St. Andrews), p. 388 (Butterworth, London, 1964).

[24] Kukhareva, L. V., Frenkel, S. Ya., Ginzburg, B. M. and Vorob'ev, I., *Biorheology*, 7: 37 (1970).

[25] Blanton, P. L. and Biggs, N. L., *J. Biomech.*, 3: 181 (1970).

[26] Rigby, B. J., Hirai, N., Spikes, J. D. and Eyring, H., *J. gen. Physiol.*, 43: 265 (1959).

[27] Bell, G. H., Cuthbertson, D. P. and Orr, J., *J. Physiol.* (Lond.), 100: 299 (1942).

[28] Burgers, J. M., *Experience and Conceptual Activity*, (M. I. T. Press, Cambridge, Mass., 1965), p. 260.

[29] Weir, J. B. de V., Bell, G. H. and Chambers, J. W., *J. Bone Jt Surg.*, 31B: 444 (1949).

[30] Dreyer, C. J. and (independently) Pearson, H. M., *Nature*, 189: 594 (1961) and 190: 1217 (1961).

[31] Bonfield, W. and Li, C. H., *J. appl. Phys.*, 38: 2450 (1967).

[32] Evans, F. G., in *Bone and Tooth* (Ed. H. J. J. Blackwood), 1st Europ. Symp., Oxford, 1963 (Pergamon Press, Oxford, 1964).

[33] Calabrisi, P. and Smith, F. C., *Naval med. Res. Inst.* NH/R-NM 001 056.02 MR51-2 (1951).

[34] Wall, J. C., Chatterji, S. and Jeffrey, J. W., *Med. biol. Engng*, 8: 171 (1970).

[35] Wood, J. L., *J. Biomech.*, 4: 1 (1971).

[36] Ommaya, A. K. and Hirsch, A. E., *J. Biomech.*, 4: 13 (1971).

[37] Holbourn, A. H. S., *Lancet*, 245: 438 (1943).

[38] Ommaya, A. K., Hirsch, A. E., Flamm, E. S. and Mahone, R. M., *Science*, 153: 211 (1966).

[39] Engin, A. E. and Wang, H-C., *J. Biomech.*, 3: 283 (1970).

[40] Ridge, M. D. and Wright, V., *Biorheology*, 2: 67 (1964).

[41] Wright, V., *Lab. Pract.*, 15: 66 (1966).

[42] Finlay, B., *J. Biomech.*, 3: 557 (1970).

[43] Harkness, M. L. R. and Harkness, R. D., *Nature*, 211: 296 (1966).

[44] Muthiah, P. L., Ramanathan, N. and Nayudamma, Y., *Biorheology*, 4: 185 (1967).

[45] Sakata, K., Parfitt, G. and Pinder, K. L., *Biorheology*, 9: 173 (1972).

[46] Veronda, D. R. and Westmann, R. A., *J. Biomech.*, 3: 110 (1970).

[47] Bull, H. B. and Gutmann, M., *J. Amer. chem. Soc.*, 66: 1253 (1944).

[48] Bull, H. B., *J. Amer. chem. Soc.*, 67: 533 (1945).

[49] Bull, H. B., *J. phys. Chem.*, 58: 101 (1954).

[50] Meyer, K. H. and Haselbach, C., *Nature*, 164: 33 (1949); also Meyer, K. H., *Nature*, 164: 34 (1949).

[51] Jagger, L. and Speakman, J. B., *Nature*, 164: 190 (1949).

[52] Treloar, L. R. G., *Nature*, 168: 521 (1951).

[53] Wright, D., *Aust. J. Sci.*, 9: 187 (1947).

[54] Copley, A. L., *Science*, 105: 341 (1947).

[55] Katchalsky, A. and Oplatka, A., in *Proc. 4th int. Congr. Rheol.*, Providence, R. I., 1963 (Eds. E. H. Lee and A. L. Copley), Part I, p. 73 (Interscience, New York, 1965),

BOOKS FOR FURTHER READING

Gaynor Evans, F., *Mechanical Properties of Bone* (Thomas, Springfield, Ill., 1973)

Needham, D. M., *The Biochemistry of Muscular Contraction* (Cambridge Univ. Press, Cambridge, 1971).

Chapter V

THE FLOW OF BLOOD, PLASMA AND SERUM

Introduction

An account of the rheology of blood, its components and vessels (which A. L. Copley has named "h(a)emorheology") has presented the most serious problem to the author of this book. The reasons are that, in the first place, about two-thirds of the papers published on biorheology are concerned with haemorheology, and secondly that at the time of writing, at least three complete books have been quite recently published on this subject, apart from the Proceedings of various Conferences (see references for further reading at the end of Chapter VI). There would be little point in merely reproducing what is to be found in these books; yet an "Introduction to Biorheology" which did not contain a fairly high proportion of haemorheology would be unbalanced. The author has therefore tried to summarize the highlights of the subject in this and the subsequent chapter without too much repetition.

In Chapter I it was pointed out that, in spite of the formal definition of "rheology", the science was generally restricted to the study of flow in so far as it is concerned with the behaviour and composition of materials, thus differentiating it from hydrodynamics, which ignores these factors. Until recently, the two disciplines were kept remarkably distinct: few hydrodynamicians knew much about rheology, nor were rheologists versed in hydrodynamics. The division is, however, somewhat artificial and recently the two subjects have tended to fuse. This has been particularly true in the case of blood. The hydrodynamics of blood flow is extremely complex in itself; yet, for practical purposes, one cannot ignore the very complicated behaviour of this material.

In these two chapters, while not ignoring the hydrodynamical aspects, we will concentrate on the rheological and, following the pattern of the rest of this book, detailed mathematical treatments will be avoided.

Jean Léonard Marie Poiseuille (1797–1869) is generally regarded as the "founder" of haemorheology, although others had certainly studied blood flow before his time. For example, the presence of a "plasmatic zone" at the vessel wall, in which there are few corpuscles, though independently noted by Poiseuille, had been already observed by Malpighi much earlier.

As is well-known, Poiseuille appreciated that blood is a complex system which flows through vessels that are neither cylindrical in cross-section, nor rigid, nor straight, nor impermeable. He therefore set out first experimentally to establish the laws of flow of simple (Newtonian) liquids such as water and alcohol through straight glass capillary tubes, as far as possible of even cylindrical bore.

Before his time, it was thought that the rate of flow of a liquid through such a tube at constant driving pressure would be proportional inversely to the length and directly to the cross-sectional area, i.e. the square of the radius. Poiseuille, working in Paris, found experimentally that the latter assumption was incorrect: rate of flow depended on the fourth power of the radius.

Unfortunately Poiseuille delayed publication for some years and, during this period, Hagen, in Germany, found the same law of flow and published the equation before Poiseuille, so it is rightly called the "Hagen–Poiseuille equation" (see Chapter II). Later, other workers showed that, making some quite simple assumptions about boundary conditions, this equation can easily be derived theoretically, based on Newton's law of flow.

It was the intention of Prof. E. C. Bingham, "founder of modern rheology", to arrange for the publication of a number of classical papers, translated from other languages, to be published in English. Unfortunately he lived to see only one such publication; some of Poiseuille's papers were translated by W. H. Herschel [1] in what proved to be the first and last of the proposed series.

The intermediate history of blood rheology will not be given here: it is to be found in the textbooks on blood. But one rather remarkable research, ahead of its time, may be quoted for its historical interest: in 1905, Rossi [2] and Fano and Rossi [3] studied the viscosity of blood, plasma (the continuous phase) and serum (the liquid left after coagulation) using a horizontal capillary viscometer to which a series of pressures could be applied. They also studied the early stages of coagulation, using the blood of dogs, horses and (strangely enough) fish.

Subsequent research on the flow of blood has faced (though not always consciously) two grave difficulties. The first is that when blood comes into contact with any surface other than the wall of a healthy blood vessel, a protein, known as the "Hageman factor" starts to polymerize and this inaugurates a chain of about a dozen reactions culminating in the formation of thrombin. Thrombin then reacts with the fibrinogen in the blood to form fibrin, which itself polymerizes to produce a clot. (In fact the process must be more complicated than this, but the above over-simplified statement will serve for the present—see Chapter VI.)

Although some surfaces, such as glass, produce clotting much more rapidly than do some others, such as silicone or stainless steel, viscometry of blood outside the body ("extracorporeal" or "extra vivum"*) has to be done very rapidly unless an anticoagulant is used. One of the essentials in the chain of the clotting reactions is the calcium ion. Salts such as sodium citrate or oxalate will remove calcium ions but they affect the ionic balance of the system, and hence its flow properties. Such anticoagulants as heparin, although so often used, have even more drastic effects. Some haemorheologists believe that a substance called EDTA (diaminoethanetetraacetic acid, disodium salt) produces no adverse effects but others disagree.

Although the present author has not had the chance of doing many experiments using an ion-exchange resin (following the suggestion of Dr. S. G. Rainsford, who has had much experience with this method), he believes that this is probably the least harmful anticoagulant rheologically. Even so, a number of platelets (see below) are destroyed.

Comparatively few rheologists (apart from the author and also A. L. Copley and L. Dintenfass) have used "native" blood, i.e. without anticoagulants, for viscosity work.

The second difficulty is that the nature of the process of shearing seems to affect the viscosity and its anomalies. More will be said in the next chapter about the different structures of thrombus formed under different conditions of shear and Scott Blair and Matchett [4, 5] have found changes in the viscosity of normal (native) human blood produced by quite gentle shearing**. Very recently Barbee [6] has found marked differences in blood viscosity (especially in relation to the effects of changes in temperature), using different types of viscometer. All this suggests extreme caution in interpreting extracorporeal rheological data in relation to what happens in vivo.

Work on "model systems", such as suspensions of plastic spheres in a liquid of the same density may help to throw some light on the hydrodynamics of blood flow, but interpretations must be made with great caution.

---

* One cannot use the term "in vitro" since glass surfaces must be avoided; "ex vivo" would mean "coming out of life".

** This phenomenon shows anomalies in most, but apparently not all cases of multiple sclerosis (see Chapter VI).

Composition of blood

Before giving a brief survey of work on extra vivum blood with, or occasionally without anticoagulants, we should perhaps provide some elementary description of blood as a colloidal system, for those who are quite unfamiliar with this branch of the subject. Blood is essentially a suspension of three types of corpuscles in a continuous medium, plasma, which is itself a non-Newtonian liquid. The composition of plasma is very complex: rheologically, the most important components are proteins.

There are three types of corpuscles suspended in the plasma: white cells (leucocytes or leukocytes) which are the largest, are not sufficiently numerous to play a major part in the rheology of the blood. They constitute the defence of the blood against invaders.

The red cells, erythrocytes, are very numerous, about a thousand times more numerous than the white cells, and morphologically very simple. They are highly flexible, with the shape approximately of an American doughnut* and compose about 40–50% of the total volume of the blood in man. Sizes and concentrations vary for different animals. (The measure of this concentration is called the "haematocrit"). The average diameter is about 7–8 microns. The erythrocytes contain red haemoglobin which transports the oxygen around the body. The rheological properties of the erythrocytes will be discussed later. In human blood, these cells tend to pile up into rod-shaped formations (rather like a pile of coins knocked sideways), called "rouleaux". Blood from some animals forms practically no rouleaux, e.g. the cow: while from others, e.g. the horse, the blood shows much more marked rouleaux formation than does human blood (Fig. V-1).

Fig. V (1). Approximate shape of an erythrocyte.

Lastly, we have the platelets. These are very small but extremely important in relation to blood coagulation both in the healing of wounds and in the formation of thrombi (see Chapter VI). Their "adhesiveness", often measured by their tendency to stick to glass, is highly significant in pathology, as are

---

* The technical name is "biconcave discoid".

also their numbers. Platelets are living cells and can project "pseudopodia", or "little arms", linking them together to form a cluster. They are about 2–3 microns in diameter and about a tenth as numerous as the erythrocytes. (For a more detailed account see Whitmore's book quoted at the end of Chapter VI.)

The viscosity of human blood varies rather widely. It is somewhat higher for men than for women and extreme values are found in certain pathological conditions. An average value (at 37°C) would be about four times that of water.

Rheological characteristics of extracorporeal blood samples

The literature on this subject is voluminous, and new articles are appearing almost daily. For the purposes of this book, a very general account must suffice.

At high shear rates (but below the level for turbulence) blood is, to all intents and purposes, Newtonian. This is not, therefore, a very interesting condition for the rheologist, since the only parameter to be measured is the viscosity. (This does not apply to flow through very narrow capillaries, which will be discussed later.) Much of this work has, therefore, a somewhat academic importance, since except in the microcirculation, for which Poiseuille's equation and its modifications are in any case inapplicable, or in extremis, very slow flow is unusual in vivo*.

An early paper on anomalies at low shear rates, using a rolling sphere viscometer, is that of Copley et al. [7], with heparinized and citrated blood, as well as other anticoagulants.

As a first approximation, blood behaved somewhat like a Bingham system, i.e. the upper part of the shear rate–stress curve was straight, but directed not to the origin, but to an intercept on the stress axis. More detailed study suggested the use of a double logarithmic plot, after first subtracting a yield value from the stress, as originally proposed for quite other materials by Herschel and Bulkley [8]. This was really prophetic—it is the present author's opinion that this equation, though not the simplest to use empirically, probably represents most closely what is happening in a viscometer with increasing shear rate (Scott Blair [9]).

Copley et al. [7] did not allow for an inertia term in their use of data from their rolling sphere viscometer. This would probably have involved quite a large correction but its absence should not affect their general conclusions.

---

* L-E. Gelin (personal communication) has pointed out its importance in transplantations.

Especially their findings concerning the rheological effects of different anti-coagulants are highly topical today.

Copley et al. [7] also verified earlier work by Trevan [10] who studied the effect of varying haematocrit on blood viscosity. At high concentrations, an equation proposed by Hatschek (under whom Trevan was working) was valid: at lower concentrations, a still simpler equation held.

Einstein had proposed the equation (for suspensions of spheres at very low concentrations)

$$\eta_{rel} - 1 = 2.5 \, \phi$$

where η_{rel} is the viscosity divided by that of the continuous phase and ϕ, the volume concentration. Trevan substituted 6.3 for Einstein's 2.5. Hatschek's equation is

$$\eta_{rel} = \frac{1}{1 - \phi^{1/3}}.$$

Copley et al. [7] also found thixotropy and a type of dilatancy (the "snow-plough effect": see Chapter IV) in blood, but the breakdown under shear, though partly recoverable (thixotropic), was not entirely so.

It is useful to have an equation to relate stress (τ) to shear rate $(\dot\gamma)$ for the flow of blood at rather low rates of shear in order to be able to express such relations in terms of a few (usually two) parameters, even if there is as yet no adequate theory for such equations.

As for many other suspensions, a plot of $\log \tau$ vs. $\log \dot\gamma$ often gives straight lines; or better, since there is evidence that blood has a yield value (τ_0), $\log (\tau - \tau_0)$. This is the Herschel and Bulkley equation; but it is not very convenient in practice, because one must first extrapolate the non-linear τ–$\dot\gamma$ curve to the τ-axis to find τ_0 before plotting double logarithms.

However, such log–log equations may have some physical meaning, as Scott Blair [11] and, more thoroughly, Cross [12] have shown.

Scott Blair [13] found that a very simple equation proposed by Casson [14] for "filled" varnishes held extremely well for the data available at the time for both bovine and human blood. This equation is very simple to use, since $\tau^{1/2}$ is found to be linear with $\dot\gamma^{1/2}$. Many later papers have vindicated the validity of this plot, except at very high and very low shear rates, and Scott Blair [15] has shown that, over the range where it is valid, the curves cannot be distinguished, either visually or statistically, from those of the Herschel and Bulkley plot. Since the "partial" theory proposed by Casson to explain his results with filled varnishes cannot be applied to cows' blood and only very tentatively to human blood (it depends on the presence of rouleaux), it is much more likely that the Herschel and Bulkley equation represents fairly closely what is happening in the blood. The "Casson plot" is now standard practice among many haemorheologists.

"Sigma" and "pinch" effects

We mentioned earlier the "plasmatic zone" in blood vessels: a region near the wall where corpuscles are few. This would imply that the R^4 term* in the Poiseuille equation would not hold and this was found to be the case, for extracorporeal blood, by Fåhraeus and Lindquist in 1931 [16]. Unknown to these authors, this same phenomenon had already been studied by rheologists not concerned with blood. The present writer's work in this field is so often misquoted by haemorheologists that it is worth while to give a brief account of it (but without references).

The first workers to report on apparently lower viscosity in very narrow capillaries than that in wider tubes were Bingham and Green, working on paints (1919), but they did not study the phenomenon quantitatively. From 1930 onwards, Schofield and the present writer made an intensive study of the phenomenon for soil and clay pastes. It was convenient to plot a $4\,V/\pi\,R^2$ against $P\,R/2\,L$ (see Equn. II-1, p. 12) since, if this equation held, one should get straight lines passing through the origin having slopes proportional to the radius. For the pastes studied, the lines were straight at higher shear rates, but the slopes were not proportional to R. Since the letter "S" was at that time used for stress, the slopes of these curves were designated by the Greek "σ" and the anomaly came to be known as the "sigma effect". Schofield and Scott Blair tested two possible explanations: one depended on the plate-like shape of clay particles and the other (perhaps combined with the first) on the postulated fall in concentration of particles near the wall. Both explanations appeared to fail, since the effect was even more marked using pastes of more nearly spherical mineral particles; and an analysis of liquid allowed to seep through a small hole in the wall of the tube showed no fall in concentration. (Probably the latter experiment was not sufficiently accurate to detect such a fall.) Consequently *no explanation was offered at the time.*

Later Dix and Scott Blair [17] suggested that, if the suspended particles were relatively large compared with the size of the capillary, it would be incorrect to derive a modification of Poiseuille's equation by integration, since the shearing layers could not be regarded as "very thin". An equation was derived by summation, assuming an arbitrary finite thickness for the shearing "layers" and was found to fit Schofield and Scott Blair's data [18]

---

* Assuming an approximately linear flow curve.

for thick soil pastes very well. It was never claimed that *all* sigma effects (by 1940 well-known for many materials) could be explained in this way. It seems most unlikely that such an explanation could apply to blood, unless in very fine capillaries—and in such cases, it seems that the effect is negative! (See Dintenfass' book listed at the end of Chapter VI.) Work on model systems has shown an even stranger effect. Segrè and Silberberg [19] found that spheres of the same density as the continuous phase moved not only away from the wall of the capillary, but also away from the centre. Experiments by Goldsmith and Mason [20] at first failed to confirm this, but in later work it was shown that the "pinch effect" (as it came to be called) is real enough within a certain range of Reynolds numbers.

These are not the ordinary "tube Reynolds numbers" (R_{et}) but "particle Reynolds numbers" (R_{ep}). Goldsmith and Mason [21] use the formula:

$$R_{ep} = \frac{R_{et}}{2} \cdot \frac{r^3}{R}$$

where r is the radius of the particle and R that of the tube. The true situation is probably even more complex than this: it has been suggested that the centrifugal motion may be caused by a type of dilatancy, even though the concentration cannot be high enough for close packing. In any case, the problem of "pinch effect" (which includes the "sigma effect") is essentially hydrodynamic.

Theoretical and experimental studies have been made on suspensions of rigid spheres, ellipsoids and other bodies and also on deformable particles. Many factors doubtless contribute to the sigma effect in blood flow. It is more difficult to explain the centrifugal behaviour.

It has also been suggested that the movement away from the wall is caused by what is known as the "Magnus effect". Even spherical particles acted on by a viscous couple will tend to rotate. This produces a force driving them across the streamlines towards the region of higher stress, i.e. away from the capillary wall. But, as Lighthill [22] has pointed out, in the case of blood corpuscles, this force must be extremely small.

We shall often refer, and not only in this chapter, to this admirable article by Sir James Lighthill. Although more concerned with hydrodynamics than with rheology, it is worthy of careful reading by biorheologists, especially those interested in the flow of air in the lungs, of blood in the circulation and of flow in the urinary tract.

First, however, we will consider a number of causes for the tendency for erythrocytes to migrate away from the wall of an artificial capillary (and also, of course, from the living blood vessel).

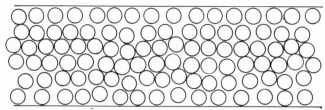

Fig. V (2). Suspension of spheres at high concentration showing slight dilution near capillary wall.

Even if the corpuscles were rigid spheres, one would predict a lower concentration at the wall. This has been explained by Vand [23] and Whitmore [24] in somewhat similar ways. It will be seen from Fig. V-2 that, in a concentrated suspension of spheres, since at the wall the centre of each sphere must be at least one radius away from the wall, whereas elsewhere there can be close packing, there must be a slightly lower concentration near the wall. But the effect is clearly not large and cannot account for the observed "sigma effects" in artificial, let alone living, capillaries.

Erythrocytes are neither spherical (normally) nor rigid. As a first approximation, we might consider what would happen if they were ellipsoid. As long ago as 1922, Jeffery [25] discussed the behaviour of ellipsoid particles. He proposed a simple modification of Einstein's equation for concentration effects, but only if inertia terms are introduced, did he find any reason to postulate movement away from the tube wall.

Many years later, Saffman [26] showed on theoretical grounds that non-spherical ("spheroid") particles showed a preferred orientation in a non-Newtonian fluid; and, in a later paper, he shows that the drag at the wall causes the particle to lag behind the fluid. But several workers have found that erythrocytes travel faster in a capillary than does the surrounding liquid, thus producing a local dilution*. This effect was first observed by Fåhraeus

---

* Dr. H. W. Thomas has kindly given me permission to quote his comments (private communication) on this apparent contradiction:

"Saffman in the paper you cited was attempting to clarify the centripetal force of hydro-dynamic origin to which a rigid particle is subject when it flows close to the tube wall. This effect is not the mechanical exclusion effect, but an additional hydrodynamic effect which is sensitive to shear rate. Saffman's starting point is that, in the neighbourhood of the rigid wall, the velocity of the centre A of the rigid sphere would be slightly less than the mean velocity of the fluid at this annular position at an infinite distance away B where the fluid flow is unperturbed, i.e. the drag of the wall causes the particle to lag behind the fluid, giving a relative velocity V. He then proceeds to examine inertial terms in the hydro-dynamic equations that could (on account of V being finite) cause a centripetal force on

56

[27] and later studied by others, including Maude and Whitmore [28] and also by Thomas [29], using radioactive isotope labels.

Many of the studies of sigma effects were made on "reconstituted blood", i.e. red cells first removed and then replaced in plasma and tested extra vivum. One must be careful in interpreting such data in terms of natural blood in vivo.

In a series of papers, of which two only will be quoted here, Bugliarello and Hayden [30, 31] studied the flow profiles (i.e. the shape of the graph of flow rate vs. distance from wall) of blood at various haematocrits and found that the profiles were "blunter" than could be accounted for by either a linear or a double logarithmic relation.

Bugliarello and Hsiao [32] offered a mathematical model for flow in smaller vessels. The same team studied the flow of spheres in channels of rectangular cross-section and found that the spheres form chains or clusters under certain conditions. The maximum concentration is sometimes part-way between the centre and the walls (Segrè and Silberberg effect).

Erythrocytes are neither rigid solids nor liquid drops—their rheological properties will be considered later. Theory and "model experiments" suggest that, had they been the former, they would tend to flow away from the wall; if the latter, they would have shunned the centre. In fact, being intermediate in properties, they do both. The unequal distribution of cells and plasma in branching systems is called "plasma skimming", a term invented by Krogh. The author is very conscious that he has omitted many important papers on this subject; but, as explained before, they are fully discussed in the recent textbooks quoted at the end of Chapter VI. It is hoped that this brief account will give the reader some idea of the nature of the phenomena.

---

an isolated particle. You will note that the final result for the centripetal force involves the terms V, but that Saffman cannot compute its value, and hence cannot provide us with a complete solution.

If the particle is subject to a centripetal force, whether it be due to Saffman's effect, due to mechanical exclusion of the particle at the wall, or due to any other effect, it will move to regions of flow closer to the axis of the tube where the mean fluid velocity is higher. In a concentrated suspension, if particles are subjected to centripetal movements, then on average they will move to annuli of higher mean fluid velocities, and hence attain a higher mean particle velocity, than if they had remained uniformly distributed across the cross-section of the tube. This is the explanation I have for the effects observed in the experiments described at the 4th International Congress of Rheology at Providence, R. I. (1963). Whitmore in his book is surely referring to the behaviour of single particles and his statement is based on the work of Segrè and Silberberg and of Goldsmith and Mason."

Viscosity of plasma and serum

We need not say much about this. Many authors have shown that plasma is non-Newtonian, the viscosity falling with increasing shear rate. There has been much unedifying argument about serum. In the author's opinion, this is because the wrong question is being asked. All "models" are "ideals" to which real materials approximate sometimes very closely, so that one can hardly object to saying that "water is a Newtonian liquid". Nevertheless, with more complex systems, the proper question is not "is it Newtonian?" but "under the conditions of our experiments, does it show significant deviations from Newtonian behaviour?"

Using admittedly heparinized blood in glass capillaries, Copley, in a specially designed viscometer, the present author and their colleagues certainly found non-Newtonian behaviour in both plasma and serum [33, 34]. They also found that, if the capillary wall was lined with a layer of fibrin, the apparent viscosity was reduced. This latter finding, which seems to support Copley's theory that blood vessels are lined with a layer of fibrin-like material in a condition of homeostasis (the "endo-endothelial layer"), has been questioned, but the authors, looking at their published graphs, find it hard to accept the criticisms of their conclusions concerning the effect of the fibrin layer on flow. It seems likely that the fibrin layer produces some "slip" at the wall of the tube. (Surface tension effects, alleged to invalidate these results, must have been negligible, since the only "surfaces" were in the wide reservoir tubes of the viscometer.)

Rheology of erythrocytes

Many of the capillaries in the vascular system have diameters smaller than that of the erythrocytes and, since the latter certainly manage to squeeze through these vessels, their own rheology is interesting.

In a most ingenious series of experiments, Katchalsky et al. [35] showed that the erythrocyte membrane could be represented at least approximately by a model of a dashpot and a spring in parallel (Kelvin model)*.

Erythrocytes were placed in an isotonic solution of NaCl and the salt concentration was reduced by dialysis so as to develop an osmotic pressure difference which eventually caused the cells to burst (haemolysis). If the concentration was reduced rapidly, a comparatively small difference of pressure would break the cells; but a gradual reduction of salt concentration

* A well-documented account of work on the rheology of erythrocytes has very recently been published by Larcan et al. [36].

allowed a much greater difference of osmotic pressure. In other words, the membrane of the erythrocyte can be stretched slowly, but it will break if subjected to a sudden stress. In the presence of 1.25% of serum albumin, the membrane becomes more resistant to osmotic pressure differences, both when slowly and when rapidly applied.

In the discussion which followed this paper, it was pointed out (by F. J. W. Roughton) that the simple process of heating erythrocytes to 48°C and then recooling to 37°C would cause them to adopt a spherical shape. It would be interesting to see whether this change of shape would affect the rheology under stress. (This change can also be effected by the addition of certain chemicals.) A. L. Copley pointed out that red cells sometimes divide into two smaller cells. On passing through the capillary, the cells are much deformed. After leaving the capillary, they regain their usual shape as two single cells.

Schmid-Schönbein, and Wells and others have shown that, under certain circumstances, erythrocytes behave like liquid drops. The subject is fully discussed in a general article by the same authors [37]. Dintenfass [38] has found that the internal viscosity of the cell varies with rate of shear.

The fine capillaries themselves are quite unlike the larger vessels. Rather inaccurately, it has been suggested that they are more like holes drilled in a solid than tubes with muscular walls.

In his review article, and in other papers, Lighthill [22] discusses, in terms of "dynamic lubrication theory", the effect of a wall layer of plasma on the rate of sliding of erythrocytes within a narrow vessel.

The rheology of the microcirculation has too vast a literature to be summarized in this book. A few papers only will be quoted here and in Chapter VI. The National and European Societies of Microcirculation have published the Proceedings (at least as abstracts) of their many conferences and these should be consulted. Some work on the rheological properties of blood vessels will be discussed in the next chapter.

Red cells that have been transformed into spheres by the action of surface-active substances ("crenated") become more rigid (see Braasch [39]).

When we discuss in vivo conditions, we shall see that there are other types of flow to be considered and some of these have been studied extra vivum. Resulting from the pumping action of the heart, there is pulsatile flow. Much work has been done on the hydrodynamics of this complicated phenomenon, but not much of this is concerned with rheology. (See Attinger [40]. What there is in the way of rheology is well summarized in this book.)

Another phenomenon is the so-called "bolus flow". Again, this is essentially a hydrodynamic phenomenon, though the rheological properties

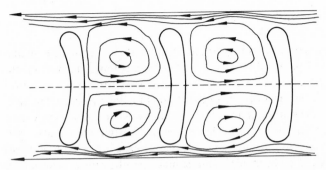

Fig. V (3). Bolus flow. Reproduced from: Rheology of the circulation by R. L. Whitmore, 1968, p. 133, Pergamon, Oxford.

of the blood doubtless play a part. The nature of bolus flow is shown in Fig. V-3.

Measurements of rheological properties in vivo

There are two principal methods for measuring the rate of flow of blood in the body. The first depends on the application of a magnetic field which sets up an electrical potential dependent on the rate of flow and which can be measured (Kolin [41]). The techniques for such measurements have naturally been much improved since 1941. A recent account of these developments is given by Wyatt [42]. It seems that the sensitivity depends on the haematocrit (except in turbulent flow, see below). But the e.m.f. (electromotive force) depends also on the distribution of the erythrocytes within the vessel and would be affected by a plasmatic zone. (For further information see Dennis and Wyatt [43].)

The second method depends on following freely diffusible radioactive indicators, generally xenon-133. This has been used to study blood flow in muscle, fat and cutaneous tissue (Lassen [44]). A number of "probe viscometers" have been described, in which blood passes directly from the vein into the capillary of the viscometer. This is not exactly "in vivo" but the method should prove useful "for monitoring the state of blood during operations with the aid of a heart–lung machine, or large infusions of plasma substitutes to patients" (quoted from Thomas et al. [45] who give references to earlier work on similar lines).

60

Turbulence in the circulation

There has been much controversy about the extent of turbulence in the large vessels in man. This is not strictly a matter for rheology, but it is worth noting here that whereas Attinger [40] maintains that turbulence is more widespread than is often supposed, Lighthill [20] is inclined to consider that there is little turbulence other than in the aorta, or where jets form when valves fail to open. Even in the aorta, he claims that turbulence occurs only during quite a small fraction of the pulsatile cycle. On the other hand, kinetic energy effects, produced by branching and changes in tube radius, would put a direct application of Poiseuille's law far off the mark in wide vessels.

The sound which the doctor hears when measuring blood pressure, when the cuff pressure is intermediate between systole and diastole ("Korotov sound"),"is a more musical sound than that produced by turbulence"(Lighthill). It is probably produced by rather regular oscillations which occur as the vessel collapses under excess external pressure.

Optical properties of plasma

In Chapter III, the significance of measurements of rheodichroism by Pfeiffer was discussed. Pfeiffer [46] pointed out that the orientation of

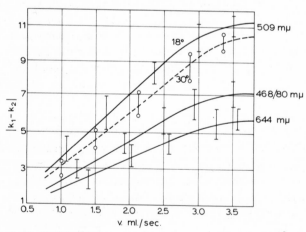

Fig. V (4). Rheodichroism distribution curves. Reproduced from: H.H. Pfeiffer, in Flow properties of blood and other biological systems (Eds A.L. Copley and G. Stainsby), 1960, p. 145, Pergamon, Oxford.

"leptones" in flowing plasma can be seen by direct visual observation. "When in motion, the plasma appears brighter and it reflects more, and absorbs less of the incident light than when at rest." But in order to find out whether the orientation agrees with a theoretical statistical distribution, a quantitative determination of rheodichroism is required. The shape of the distribution curves is shown in Fig. V-4.

In general, the dichroism increases proportionally to the velocity over a fairly wide range of shear rate; but, strangely enough, there is some slight dichroism before any shear thinning can be observed. If the speed is increased abruptly, there is sometimes a fall in dichroism, probably caused by an irreversible breaking of bonds between the leptones. The author concludes that the leptones in plasma are probably arranged in "cybotactic groups".

REFERENCES

[1] Poiseuille, J. L. M., transl. W. H. Herschel as *Rheological Memoirs*, Vol. 1 (1) (Publ. Lafayette College, Easton, Pa., 1940).
[2] Rossi, G., *Arch. Fisiol.*, 2: 246 (1905), 2: 272 (1905) and 2: 599 (1905).
[3] Fano, G. and Rossi G., *Arch. Fisiol.*, 2: 589 (1905).
[4] Scott Blair, G. W. and Matchett, R. H., *Rheol. Acta*, 10: 49 (1971).
[5] Scott Blair, G. W. and Matchett, R. H., *J. Neurol. Neurosurg. Psychiat.*, 35: 730 (1972).
[6] Barbee, J. H., *Biorheology*, 10: 321 (1973).
[7] Copley, A. L., Krchma, L. C. and Whitney, M. E., *J. gen. Physiol.*, 26: 49 (1942).
[8] Herschel, W. H. and Bulkley, R., *Proc. Amer. Soc. Test Materials*, 26: 621 (1926).
[9] Scott Blair, G. W., in *Hemorheology, Proc. 1st int. Conf.*, Reykjavik, 1966 (Ed. A. L. Copley), p. 345 (Pergamon Press, Oxford, 1968).
[10] Trevan, J. W., *Biochem. J.*, 12: 60 (1918).
[11] Scott Blair, G. W., *Rheol. Acta*, 6: 201 (1967).
[12] Cross, M. M., *Rheol. Acta*, 10: 368 (1971).
[13] Scott Blair, G. W., *Nature*, 183: 613 (1959).
[14] Casson, N., *Brit. Soc. Rheol. Bull.*, No. 52: 5 (1957) and in *Rheology of Disperse Systems* (Ed. C. C. Mill), p. 84 (Pergamon Press, London, 1959).
[15] Scott Blair, G. W., *Rheol. Acta*, 5: 184 (1966).
[16] Fåhraeus, R. and Lindquist, T., *Amer. J. Physiol.*, 96: 562 (1931).
[17] Dix, Y. F. and Scott Blair, G. W., *J. appl. Phys.*, 11: 574 (1940).
[18] Schofield, R. K. and Scott Blair, G. W., *Proc. roy. Soc. A*, 138: 707 (1932); 139: 557 (1933); 141: 72 (1933); 160: 87 (1937).
[19] Segrè, G. and Silberberg, A., *Nature*, 189: 209 (1961).
[20] Goldsmith, H. L. and Mason, S. G., *2nd Europ. Conf. Microcirculation*, Pavia, 1962, p. 462.
[21] Goldsmith, H. L. and Mason, S. G., in *Rheology: Theory and Applications*, Vol. 4 (Ed. F. R. Eirich), p. 86 (Academic Press, New York, 1967).
[22] Lighthill, M. J., *J. Fluid Mech.*, 52: 475 (1972).

[23] Vand, V., *J. phys. Colloid Chem.*, 52: 277 (1948), 2: 300 (1948) and 2: 314 (1948).
[24] Whitmore, R. L., Chapter III in *Rheology of Disperse Systems* (Ed. C. C. Mill) (Pergamon Press, London, 1959).
[25] Jeffery, G. B., *Proc. roy. Soc. A*, 102: 161 (1922).
[26] Saffman, P. G., *J. Fluid Mech.*, 1: 540 (1956) and 22: 385 (1965).
[27] Fåhraeus, R., *Physiol. Rev.*, 9: 241 (1922).
[28] Maude, A. D. and Whitmore, R. L., *J. appl. Physiol.*, 12: 105 (1958).
[29] Thomas, H. W., *Biorheology*, 3: 36 (1965).
[30] Bugliarello, G. and Hayden, J. W., *Science*, 138: 981 (1962).
[31] Bugliarello, G. and Hayden, J. W., *Trans. Soc. Rheol.*, 7: 209 (1963).
[32] Bugliarello, G. and Hsiao, G. C., *Biorheology*, 7: 5 (1970).
[33] Copley, A. L., Scott Blair, G. W., Glover, F. A. and Thorley, R. S., *Kolloid Z.*, 168: 101 (1960).
[34] Copley, A. L. and Scott Blair, G. W., *Rheol. Acta*, 1: 170 (1958) and 1: 655 (1961).
[35] Katchalsky, A., Kedem, O., Klibansky, C., and de Vries, A., in *Flow Properties of Blood and Other Biological Systems* (Eds. A. L. Copley and G. Stainsby) (Pergamon Press, Oxford, 1960).
[36] Larcan, A., Stoltz, J. F. and Vigneron, C., *Cah. Gr. Tranç. Etud. Rhéol.*, 3 (1): 10 (1973).
[37] Schmid-Schönbein, M. and Wells, R. E., *Biorheology*, 7: 227 (1971); *Ergebn. Physiol.*, 63: 145 (1971).
[38] Dintenfass, L., *Nature*, 219: 956 (1968).
[39] Braasch, D., *Proc. 6th int. Congr. Microcirculation* (Eds. J. Ditzel and D. H. Lewis), Aalborg (Abstr.) (Karger, Basel, 1970).
[40] Attinger, E. O., *Pulsatile Blood Flow* (McGraw-Hill, New York, 1964).
[41] Kolin, A., *Proc. Soc. exp. Biol. Med.*, 35: 53 (1936–7) and 46: 235 (1941).
[42] Wyatt, D. G., in *Theoretical and Clinical Hemorheology* (Eds. H. H. Hartert and A. L. Copley) (Springer-Verlag, Berlin–Heidelberg, 1971).
[43] Dennis, J. and Wyatt, D. C., *Circulat. Res.*, 24: 875 (1969).
[44] Lassen, N. A., *Proc. 6th int. Congr. Microcirculation*, Aalborg (Abstr.) (Karger, Basel, 1970).
[45] Thomas, H. W., James, D. E. and Barnes, H. A., *J. Phys. E*, Ser. II, 1: 834 (1968).
[46] Pfeiffer, W., in *Flow Properties of Blood and Other Biological Systems* (Eds. A. L. Copley and G. Stainsby), p. 145 (Pergamon Press, Oxford, 1960).

Chapter VI

BLOOD COAGULATION: HAEMORHEOLOGY AND PATHOLOGY; RHEOLOGY OF BLOOD VESSELS

Coagulation (general principles)

The rheology of blood coagulation is dealt with very fully in the book by Dintenfass (see end of this chapter), and this source, as well as many of Dintenfass' numerous scientific papers, are gratefully acknowledged in the writing of this chapter. Nevertheless, although some repetition is inevitable, there are various aspects of the subject which will be quoted from other sources.

In the last chapter, the cascade of reactions, starting with the activation of the Hageman factor as a result of contact with a foreign surface, was mentioned, but it was pointed out that this was not the only mechanism of coagulation. This cascade process is known as the "intrinsic process", but there is also a so-called "extrinsic process" activated by enzymes in tissue juices from damaged tissue.

For this reason, some haematologists (and haemorheologists) always discard the first small volume of blood taken from a vein for test purposes (see Lalich and Copley [1]). But, apart from the extrinsic system, even the intrinsic system is far more complex than the simple classical "cascade theory" would suggest. Happily, the rheologist is only indirectly concerned with all these biochemical processes: his aim is to follow the rheological changes that take place before, during, and after coagulation. Nevertheless, it would be worth while to read Biggs and Macfarlane's classical book on this subject (see end of this chapter).

The reader is warned that the numbers given to the separate factors can be misleading to the uninitiated. Broadly, they number backwards: the Hageman factor is numbered XII (to add to the confusion, a XIII has now also been added); VI proved to be a mistake and is omitted, and fibrinogen, which reacts with thrombin to form the fibrin which polymerizes, is numbered I. Factor X comes in the wrong place and, although most of the "factors" are proenzymes, which produce enzymes to interact with the neighbouring factor, one of them is simply calcium, the need for which occurs at more than one stage in the process.

An almost total absence of any one of these factors will make coagulation extremely slow. The concentrations are generally described as proportions of the normal average; thus a factor VIII level of 50 would mean half the normal average, and blood at this level would not normally be very slow to clot.

Fortunately only two factors are found to be deficient, except very rarely, and deficiencies in a few of the others have been claimed only in one or two cases throughout the world. Factor VIII (haemophilia) and IX deficiency (Christmas disease) will be discussed later; since, although fortunately fairly rare, they are those most frequently found and they can be treated. One of the difficulties in all this work is that so little is known about the chemical constitution of these proenzymes and enzymes and the methods for determining them (fortunately not relevant in this book) are indirect.

Quite crude, but useful measures of "clotting time" are commonly made by direct observation of blood in a glass test tube held at 37°C.

During the clotting process, blood changes from being a nearly Newtonian liquid (at all but very low shear rates) to become a solid "clot", which nevertheless still retains some "imaginary" component of the complex modulus.

In many haematological laboratories, this process is measured by means of a commercially available instrument called the "thrombelastograph". Before discussing this, however, some early work by the author, using a U-tube rheometer, will be described.

The U-tube rheometer

This apparatus has the advantage of extreme simplicity but the disadvantage that, in its original form, it requires 50 ml of blood, which makes it unsuitable for much of the work on human blood. The tests now to be described were done on blood taken from the jugular vein of the cow (Scott Blair and Burnett [2] and Scott Blair [3]).

The blood is placed in a silicone-coated U-tube, the surfaces being covered with a neutral liquid such as a silicone or kerosene. The U-tube is connected to a four-way tap so that air pressure may be applied to either arm. The pressure is applied by squeezing a rubber tube through a small double-roller "mangle" and is measured on a suitable manometer. The arm of the U-tube not under pressure is attached to a horizontal capillary tube in which is placed a small drop of coloured alcohol containing a surfactant, so that its surface tension is negligible. When a pressure is applied, there is a depression in the meniscus in the "pressure" arm and a corresponding rise in the other arm. This deformation can easily be magnified a hundredfold by the movement of the drop of alcohol along the capillary. The U-tube is kept in a water-bath at constant temperature. (A much smaller version of this apparatus was also described in the literature to be suitable for small samples of human blood but the accuracy was not very good.)

Before the blood starts to coagulate, applying a very small pressure will produce a large deformation simply as a result of static head. This is, of course, ignored. Most of the interesting results were obtained when the blood was in a fairly stable condition: i.e. fully clotted but not yet softening appreciably.

Creep tests at constant stress showed remarkably linear curves plotting displacement vs. time (Fig. VI-1) and pressure–displacement curves showed that the viscous part was remarkably linear and the elastic part, nearly so (Fig. VI-2). If the pressure was raised rapidly until a certain strain was reached, and this strain was then maintained by gradually reducing the pressure, relaxation curves could be obtained, which, plotted as log stress vs. time, gave excellent straight lines until a stress of well below $1/e$ of the original stress was reached (see Fig. VI-3). This means that the fully coagulated blood behaves very much like a Maxwell model: one spring and one dashpot in series. It is also interesting that the immediate elastic response to stress is not diminished by "static fatigue" during creep, since the immediate recovery, when stress is released, is as big as the initial response (see Fig. VI-1).

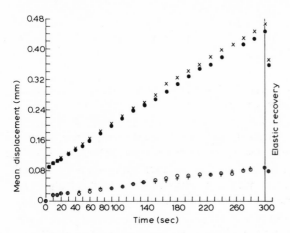

Fig. VI (1). Flow curves for bovine blood. Reproduced from: Scott Blair, G.W. and Burnett, J., Kolloid Z., 168: 98 (1960).

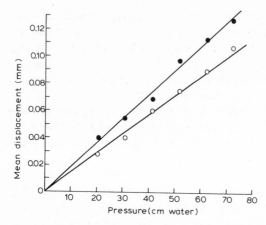

Fig. VI (2). Elastic and viscous displacements of bovine blood. Reproduced from: Scott Blair, G.W. and Burnett, J., Kolloid Z., 168: 98 (1960).

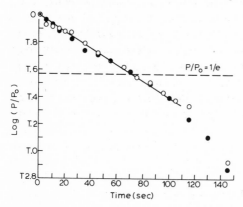

Fig. VI (3). Duplicate relaxation curves for bovine blood. Reproduced from: Scott Blair, G.W. and Burnett, J., Kolloid Z., 168: 98 (1960).

In a further experiment, alternate tests were made using direct loading (and so calculating viscosity and shear modulus) and relaxation. For a Maxwell model, the relaxation time should be equal to the viscosity divided by the shear modulus. Relaxation times measured directly and calculated by this ratio were found to agree well once the clot was fairly firm, but not during the earlier stages of coagulation.

In order to take blood from the vein, the cow's head had to be tied to a post. Though cows never seemed to object to the actual taking of the sample, tying the head sometimes disturbed them considerably and sometimes hardly at all. The "degree of disturbance" was estimated on a five-point scale and was found to correlate quite highly with the rate of coagulation of the blood.

Studies have been made on the rate of coagulation of blood in nervous patients and some such effects (though much less marked) appear to have been observed. (Of course the patients can object only to the taking of the sample from the antecubital vein: they do not have to have their heads immobilized!)

In a paper in Japanese (but with an excellent English summary) Fukada and Kaibara [4] study both steady-state viscosity and dynamic visco-elasticity of blood during coagulation using a coaxial cylinder rheometer. Within the viscous range, a Casson plot gives excellent straight lines. When coagulation is nearing completion, the "real" part of the complex modulus is approximately linear with time (i.e. between about 12 and 40 min). Relations with plasma concentrations are also discussed.

The writer is aware of very little further work on the fundamental rheology of fully coagulated blood; but much has been published on the changes that take place during the coagulation process. Only a few "representative" examples can be quoted here.

Two best instruments for such studies are, undoubtedly, the rheogoniometer invented by K. Weissenberg (see Chapter II) and widely used for such studies by Copley and his colleagues, and the cone-plate, or better cone-cone, as used by Dintenfass.

Copley has written many papers on this subject, especially related to surface problems*: we have space to quote only three [5–7]. The first of these papers is concerned with fibrin gels (not whole blood). Sinusoidal strain (induced) and stress (resultant) curves are shown which give a zero phase angle, i.e. the reaction is entirely elastic. When the rigidity modulus is plotted against time (up to 200 min), there are found to be two marked breaks in the curve; i.e. regions in which the modulus, increasing elsewhere, remains almost constant. These lie around 20 min and onwards from 49 min. These "kinks" are "discussed in relation to the roles of fibrinopeptides A and B as well as other factors which may contribute to the gel structures".

The second paper quoted is concerned with the use of guard-rings in the apparatus in studying plasma and serum systems, but it bears on the question of coagulation, since Copley draws conclusions from it concerning the initiation of thrombosis "based on the formation of polymolecular layers of fibrinogen and other plasma proteins leading to obstructions of the affected blood vessel and impairment of the circulation".

The third paper is also concerned with surface layers of fibrinogen and other plasma proteins and is again directed at a possible cause of thrombosis.

*See APPENDIX p. 199

(Although optical, and not rheological methods were used, mention should perhaps be made of the pioneer work of Ferry [8] and Lorand [9] on the polymerization of fibrin.)

The work of Dintenfass and his colleagues is very fully described and many references are given in his book (see the end of this chapter). Nevertheless, some brief account should be given here.

It has long been known that the morphology of blood clots is greatly dependent on the way in which they are allowed to form (reported as early as 1882 by Bizzozero). A study of the effects of varying the shear rate (using a cone-cone viscometer) during coagulation was published by Rozenberg and Dintenfass [10, 11]. In his book, Dintenfass describes many experiments showing the fall in viscosity of coagulating blood with increase in shear rate (which, unlike almost all other rheologists, he calls "thixotropy"*, see Chapter I) and also the subsequent breakdown of the clots. The effect of increased shear rate on viscosity is well shown in a figure from Dintenfass' book (Fig. VI-4). The visco-elasticity of the final clot is also markedly affected by the rate of shear during the clotting.

There are many empirical standardized tests for "clotting time" of blood. A very useful test (and less arbitrary than most) was proposed by Janes and Thomas [12], who found that the moment when blood starts to "climb" a vertical rod slowly rotated in it gave a useful "clotting time" (although in fact the authors used mixtures of thrombin and fibrinogen and not native blood).

Dintenfass also shows, in his book, pictures illustrating the marked differences in structure of thrombi formed under different conditions of shear. At high shear rates, white thrombi are formed, containing large masses of platelets: at very low rates, red clots preponderate, the erythrocytes being embedded in a mass of fibrin.

Dintenfass ascribes the differences in the morphology of the clots to four main factors: "(1) the progressively increasing diffusion of the coagulation factors and profactors . . . (2) the progressive orientation of the fibrin network . . . (3) the progressive disaggregation of red cell aggregates and a progressive escape of the red cells from the fibrin network; (4) progressive aggregation of platelets . . . ". Dintenfass introduces a quantitative measure of the degree of aggregation of platelets and this is related by a double logarithmic equation to the velocity gradient. Dintenfass' work on pathological blood will be discussed later.

---

* H. Freundlich, who invented the word "thixotropy", explicitly repudiated this use of the term. When there is no observable time effect, the standard term is "shear thinning". Dintenfass points out that the term "observable" is arbitrary: yes, but although some marbles "flow" in a century or so we still call marble "a solid"!

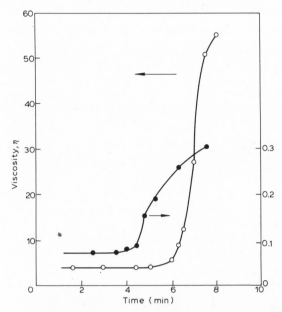

Fig. VI (4). Changes in blood viscosity during the initial stages of clotting. The viscosity, η (poises), was followed at shear rates of 0.18 $\sec^{-1}$ (circles) and 56.8 $\sec^{-1}$ (dots), and is plotted against the time (min.). Reproduced from: L. Dintenfass, Blood Microrheology, 1971, p. 164, Butterworths, London.

The thrombelastograph

In common practice, haematologists seldom use U-tube or cone-cone rheometers. The commercially available "thrombelastograph", invented by Prof. H. H. Hartert of Heidelberg, has been widely used. This instrument, of which a picture is shown in Fig. VI-5, consists of three very small stainless steel containers (each requiring only about 0.3 ml), in which are hung three steel cylinders supported by torsion strips; thus forming, in effect, three independent coaxial cylinder rheometers. The containers are oscillated by means of a cam to give a sinusoidal motion* and the corresponding sinusoidal stress curve (since blood behaves like a Maxwell model) is recorded photographically, as well as being observable visually on a small screen. A typical "thrombelastogram" is shown in Fig. VI-6.

---

* In the original technique, a 1-second pause was given at the end of each swing. It was mistakenly believed that this eliminated measurements of viscosity. But the "pauses" are easily eliminated.

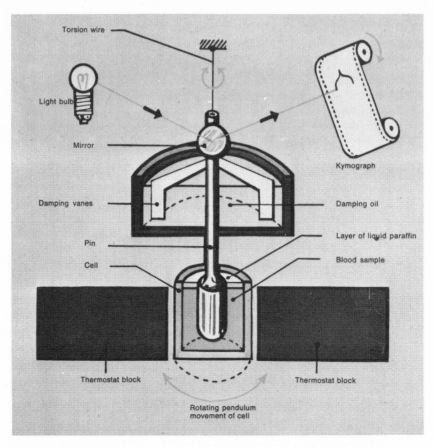

Fig. VI (5). Thromb-Elastograph (H. Hartert). Patent: Hellige, Freiburg, Germany.

The width of the double-swing of the suspended cylinder gives a measure of the initially increasing complex modulus of the coagulating blood, but it is not possible, with this instrument, to measure the phase angle and so to distinguish the real from the imaginary parts of the modulus. The stress is, of course, given directly by the torque on the inner cylinder, but the strain depends on the difference in angular motion of the two cylinders. Hartert gives a simple formula to allow for this in calculating the modulus.

It was originally stated that the instrument "did not measure viscosity" —the swinging cylinders are "damped" with oil and there was the pause at the end of each swing. In fact, the viscosity of the blood before any coagulation takes place is not measured simply because the torsion strips are too rigid to record it. But, if highly viscous Newtonian fluids are tested in the instrument, thrombelastograms (of a very unusual shape) are in fact produced (see Scott Blair and Burnett [13]). The pause at the end of each swing allows for some stress relaxation, but since the relaxation time of clotted blood is of the order of 1 min, this must be far from complete, even allowing for the fact that the fastest relaxation occurs at the start. Moreover, both stress and strain are relaxing, so that there can be no "elimination" of viscosity. The idea probably arose through lack of appreciation of the significance of the imaginary part of the complex modulus, which is a measure (at any given frequency) of the viscosity.

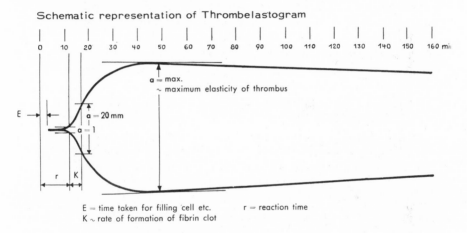

Fig. VI (6). A typical thrombelastogram (H. Hartert). Patent: Hellige, Freiburg, Germany.

Fig. VI-6 shows a number of arbitrary parameters by which thrombelastograms were originally designated. In fact all these, except the maximum width, are quite empirical.

The present writer has tested perhaps about a hundred samples of blood —bovine, human controls and pathological—and has found in every single case (except when coagulation is so slow that it is impossible to define the moment when it starts) that thrombelastograms can be expressed by a very simple equation and may be described in terms of three parameters: $G_\infty^\star$, the value of the complex modulus if the equation held to infinity (which it does not because of "softening" to be discussed later); $G_{\max}^\star$, the effective maximum modulus which does not appear in the equation; and τ, the time taken for $G^\star$ to reach a value of $G^\star_\infty/e$. The equation is:

$$G^\star = G^\star_\infty\, e^{-\tau/t} \qquad\qquad\qquad \text{(Equn. VI-1)}$$

Several papers were published about the significance of this equation by the writer. He now feels that these were somewhat premature. This is a "basic" equation, which means that, like all "model equations", it is an ideal based on certain simplifying assumptions. If these do not hold perfectly, the equation would have to be modified accordingly. It seems likely that in the case of coagulating milk (with rennet) some such modifications are needed, though the model equation was originally proposed for milk (W. Tuszynski, private communication). This is perhaps because, whereas with blood, the monomer, fibrin, is being produced all the time during coagulation; with casein the substrate concentrate remains substantially constant. The inventor of the thrombelastograph, Prof. H. H. Hartert, has very kindly expressed his approval of the comments that the writer has made and the changes he has proposed in the interpretation of the thrombelastograms.

Equn. VI-1 can be easily derived if we make a few very simple assumptions: (1) It is assumed that any polymerization process must be represented by a sigmoid curve. At the start, there must be an acceleration, partly because when two molecules join at one point they will be near enough to make junctions very soon at other points. Towards the end, the number of free junction points must run out, and the process will slow down. (2) If n is the number of junctions per unit volume at any time t, then the rate at which $G^\star$ will increase with n must be a function of either n or $G^\star$. It can hardly be a function of n, because an evenly dispersed arrangement of bonds would clearly give a higher value of $G^\star$ than would an uneven distribution, consisting of, say, groups of molecules with many bonds within each group but with the same value of n. Hence we have $dG^\star/dn = f_1\, G^\star$. (3) We assume that dn/dt is some inverse function of time as the free junction points become more scarce, or $dn/dt = f_2\,(1/t)$. Combining these equations we get,

$$dG^\star/dt = f_3\, G^\star/t \qquad\qquad\qquad \text{(Equn. VI-2)}$$

(4) The function f_3 must be a numeric, to balance the dimensions, but it cannot be just a simple number, because the equation would then become a power equation which would give a curve sloping upwards or downwards all the way, depending on whether f_3 was greater or less than 1. Hence, f_3 must start by being very large (steep upward curvature) and progressively diminish to become very small. (5) The simplest assumption here is that f_3 will be proportional to $1/t$, or

$$\frac{\mathrm{d}G^\star}{\mathrm{d}t} = \tau\, G^\star\, t^{-2} \qquad\qquad \text{(Equn. VI-3)}$$

where τ balances the dimensions and is a characteristic of the system. Integration of this equation gives Equn. VI-1, since the integration constant must correspond to the value of $G^\star$ when $t = \infty$, or $\ln (G^\star/G_\infty^\star) = t/\tau$ which is the same as Equn. VI-1.

The equation is easy to use, because graph paper can be prepared with a reciprocal scale on one ordinate and a logarithmic scale on the other. Provided that the time zero is very carefully determined, this plot gives remarkably good straight lines for blood (see Scott Blair and Matchett [14]).

Some typical curves from this paper are shown in Fig. VI-7.

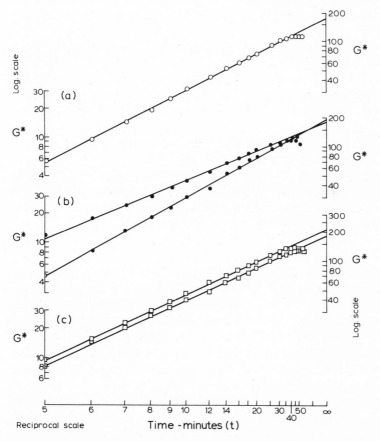

Fig. VI (7). Some typical thrombelastogram curves plotted according to the equation log G^* plotted against $1/t$. Reproduced from: Scott Blair, G.W. and Matchett, Haemostasis, 1: 93 (1972).

It must be pointed out, however, that curves plotted on non-linear paper can sometimes give a misleading impression of linearity. The equation is based on the simplest assumptions and it may well be that further work will show that some small modifying terms are needed.

Thrombosis in vivo will be discussed later: here it should be pointed out that as early as 1945, Copley [15] had proposed the existence of a third phase in coagulation, differing from fibrin monomer polymerization.

It will be seen in Fig. VI-7 that, whereas $G_\infty^\star$ can be determined by extrapolating the straight lines to infinity on the reciprocal time scale, there is, in fact, a maximum value of $G^\star$. This is a third parameter that does not appear in the equation. After this value is reached, the curves start to fall. It is essential that blood clots should eventually dissolve or disintegrate if wounds are to heal and if minor thromboses are to be dissipated.

This is affected by two, and possibly three independent processes. The first involves a complex series of, mainly, enzyme reactions and is known as "fibrinolysis". With healthy human blood this process takes place very slowly: when we come to discuss pathological blood we shall see that in some conditions it can be very much more rapid. There is also syneresis—the expression of liquid from the fibrin gel, similar to that found in most other gels such as gelatine; and finally "retraction". There are differences of opinion as to whether those last two phenomena are really distinct. They are observed separately in the thrombelastograms, the former showing simply a narrowing of the diagram and the latter, complete irregularities caused by the formation of pockets of serum on the surface of the steel cylinders. (For further discussion see Budtz-Olsen [16] and Rubenstein [17]). Various empirical devices have been proposed for the measurement of retraction.

The former author measured the forces involved in retraction and concluded that they were far smaller than those which occur in syneresis. He believed that retraction is caused by the action of the platelets which develop long strands of cytoplasm and form a network from which serum is squeezed by very small forces. The latter author, however, published experimental evidence which he claimed to show that the forces are considerable. The effect of platelets in clot retraction was also discussed by Hartert [18]. At the same Conference, the present author quoted data obtained by Burnett and Scott Blair [19], using a home-made instrument working on much the same principles as the thrombelastograph, in which they showed that the softening process (for bovine and human blood) followed an exponential equation. In the case of bovine blood, only a fraction of the rigidity is destroyed; with human blood, complete liquefaction is finally reached. Normal bovine blood does not appear to show fibrinolysis.

The system of reactions involved in fibrinolysis (the "plasminogen–plasmin system") is probably as complex as the coagulation system and is even less fully understood. Yet the therapeutic use of fibrinolytic agents rather than anticoagulants for the treatment and prevention of thrombosis is now widespread. The subject, insofar as it was understood in 1962, is well summarized in an article by Douglas [20]. Urokinase and streptokinase induce thrombolysis in the human and ϵ-aminocaproic acid inhibits it. Scott Blair and Burnett [21] confirmed similar effects with bovine blood, except that streptokinase is not effective. Various other substances were used: methyl violet inhibits softening but methyl blue has no effect. Morphine increases softening and also changes the shape of the curve. ϵ-Aminocaproic acid has no effect on bovine blood clots.

Tensile strength of blood clots

Several workers have measured the tensile strength of blood clots both in health and disease.

Ferry and Morrison [22] and Dees and Fox [23] worked on pure fibrin: Kristensen [24] and Rubenstein [25], on plasma and Macfarlane and Tomlinson [26] on whole blood. The importance of this work lies chiefly in changes that take place under pathological conditions. The experimental methods generally consist of straightforward stretching to rupture. But Macfarlane et al. [27] describe an apparatus in which a blood clot is first cast into the shape of a dumb-bell. This test piece is broken by means of two pendulums, the only connection between which is the strand of clot. The traction is applied from a geared motor. The measuring pendulum is held stationary at its highest point and the distance it has travelled is measured, from which the tension at rupture can be calculated. This method ensures that the tension is always along the axis of the clot.

Coherence of erythrocytes

Erythrocytes are all charged negatively, so that if no other forces were involved, one might suppose that they would remain discretely apart. However, it has long been known that in some species and under certain conditions, they do clump together. With anticoagulated blood, this clumping causes a sedimentation. Dintenfass claims that this was noted as early as 1772 by Hewson and by others in 1826, 1840 and 1897.

But the first serious study of this effect was made by Fåhraeus [28] in 1918 who noticed that blood from pregnant women sedimented faster than that of non-pregnant women and believed that he might have found a convenient test for pregnancy. However, he soon afterwards found the same more rapid sedimentation in some male patients! It was then established that the "erythrocyte sedimentation rate" (ESR) served as a good indication that all was far from well, with both men and non-pregnant women, but that quite a variety of conditions, many of them serious, increased the ESR. For many years, this has been a standard "trouble shooter" in hospitals all over the world; but the test must be very carefully done and agreement between different testers has not always been good. It is interesting that Harkness [29] has found that an abnormally high viscosity of plasma, as measured in a specially designed capillary viscometer, gives much the same results as the ESR, with less personal error. Thinking only in terms of Stokes' law, one would have supposed that a more viscous plasma would have slowed down the sedimentation rate of the red cells. Clearly there is much more involved than this: there must be quite characteristic changes in the plasma proteins.

It has already been mentioned that human erythrocytes form "rouleaux" whereas bovine blood does not, nor does it sediment appreciably. A vigorous controversy arose some years ago between Prof. Fåhraeus in Sweden and Prof. Knisely in S. Carolina, U.S.A., concerning, mainly, the blood of horses. Knisely had been studying the "sludging", or clumping together of erythrocytes in (he believed) *all* pathological conditions and claimed that, in perfect health, the red cells in vivo remained completely detached from one another. Fåhraeus claimed to find evidence of "clumping" in the blood of horses that could in no way be other than healthy. Knisely, while admitting that equine blood showed high ESR values, claimed that in vivo there was no "clumping" unless there was some injury to tissue. He showed that some substance could be exuded from damaged muscle that could produce clumping. Knisely and his colleagues published many papers on this subject: an attempt has been made to select for quotation here the three most representative [30–32]. The last of these is a general review on Knisely's work. For a later contribution to the discussion see Fåhraeus [33].

Dextran

There is a vast literature on the complex effects of the addition of dextran of various molecular weights to blood. Only a few key papers can be quoted here. A pioneer in this field is Gelin [34, 35]. Very high viscosity dextran increases the viscosity of blood in vivo and also the aggregation of the blood cells; whereas low viscosity dextran serves as a "plasma expander" and reduces viscosity and aggregation. This latter effect was also stressed in a paper given at the same Congress as that of Gelin, by Meiselman and Merrill [36]. Dextrans are carbohydrates prepared from glucose, generally by means of bacteria. In clinical practice, two grades are in general use: the "flow improver type" (low molecular weight dextran: LMWD) which has an average molecular weight of about 40,000 and the expander "type" (high molecular weight dextran: HMWD) with a molecular weight of about 80,000 or more. One effect of low molecular weight dextran is to reduce mutual red-cell interaction but this is not believed to be the main cause of the improvement in flow. This depends rather on the reduction in the concentrations of both red cells and plasma proteins. These authors compare their treated and control samples by using a Casson plot of square roots of stress and shear rate, which give excellent straight lines over the range covered.

Brooks and Seaman [37] added low molecular weight dextrans to concentrated suspensions of erythrocytes in isotonic saline and this produced an increase in the electrostatic repulsion between the cells which lowers both viscous anomaly and viscosity itself. Both optical and rheological (rheogoniometer) methods were used. The effects of increasing concentration of cells in increasing anomalies were also reduced by the dextran. The authors ascribe these effects to adsorption of dextran molecules at the surface of the cells.

Wells [38] likewise studied the effects of additions of dextran (molecular weight 40,000) on erythrocyte aggregations. Aggregation was produced "by the addition of fibrinogen to freshly drawn blood. One part of dextran added to four parts of blood reduced viscosity and hæmatocrit by 20% but did not influence erythrocyte aggregation, when in solution prior to the addition of fibrinogen, or when added after the addition of fibrinogen." In discussing Wells' paper, Ehrly, quoting a publication of his own [39], said that he had observed no disaggregation of rouleaux in vitro, but that there was disaggregation in vivo "induced mainly by haemodilution". Dextran (molecular weight 40,000) reduces the ESR without disaggregation of rouleaux.

Pathological conditions

The author is not a medical man and his summary of the effects of various pathological conditions must be written from the rheological point of view. Dintenfass, though also not medical, has much information in his book, as have also the two French authors quoted at the end of this chapter. (Monsieur Larcan is a Professor in the Faculty of Medicine at Nancy.)

ESR and its parallel, plasma viscosity, like the viscosity of blood itself, can give the physician evidence that all is far from satisfactory with the patient, but the former two tests show abnormalities in so many diseases that they can do little more than indicate the need for further investigations.

Dintenfass discusses what he calls the "high viscosity syndrome". This again may be associated with many diseases, though it is typically characteristic of a few, such as Raynaud's disease. It has been claimed that active multiple sclerosis (MS) patients show high blood viscosity (Proewig [40]) but the author has examined 45 samples of blood from 27 patients and has found that, while some have a very high viscosity, in others the viscosity is exceptionally low. He is under the *impression* (not statistically verifiable with such small numbers) that the range of blood viscosity is probably greater for active MS patients than for normal people.

Although this work was unfortunately left unfinished, a preliminary note was published and already referred to (Scott Blair and Matchett [41]) in which experiments with a specially designed capillary viscometer showed a definite fall in viscosity following (and probably resulting from the first shearing) in all cases (including a number of patients with other neurological diseases); whereas in the majority (but not all: 19 out of 27) of MS patients, there was no such fall. Hitherto, the only well-attested effect of MS on blood has been an increase in platelet adhesiveness and this also occurs with a number of other diseases (see Sanders et al. [42]).

Dintenfass, in his book and in his many papers quoted in the bibliography, discusses the effects of a number of diseases, not so much on the blood viscosity itself, as on the change in viscosity with shear rate. It seems likely that the yield value and the viscosity at very low shear rates are more affected by many pathological conditions than is the viscosity when the blood is flowing fast. Dintenfass shows viscosity–shear rate curves for native blood from eight haemophilia patients but the conclusions to be drawn are not very clear. Some, by no means all, of the curves (plotted log–log) show sharp "kinks" at the shear rate at which it is thought that disaggregation of the red cells is taking place. There is no doubt that Knisely is correct in stressing that pathological conditions are associated with a special type of cell aggregation but, whether observed directly under the microscope or indirectly through the changes produced in anomalous viscosity, very little specific information can be obtained about the nature of the disease. Viscosity (at high shear rates) is, quite naturally, associated with haematocrit, but the precise relationship—exponential, or power law—is in dispute. Other conditions involving excess or deficiency of leucocytes or platelets likewise affect viscosity.

It is of course well-known that diabetes often produces impaired blood flow. The reasons for this have been studied, especially by Ditzel [43, 44] who ascribes it to many causes, including aggregation of erythrocytes, changes in the vessel wall, and levels of fibrinogen and of various globulin proteins. As with MS there is also an increased adhesiveness of platelets. Abnormally high levels of certain clotting factors are also alleged by various authors to be associated with high blood viscosity.

Not unnaturally, the disease which has aroused the greatest interest has been thrombosis and other ischemic conditions, because of their now alarming prevalence. Here again, it is true that patients who have had a thrombosis tend to show the "high viscosity syndrome" but, from what has already been said, it is evident that this fact is not of much help in warning those "at risk" for this type of disease. Hypertension also is sometimes, but of course by no means always, a warning of such attacks.

We have given here a brief and perhaps rather arbitrary selection of what is known about the significance of Dintenfass' "high viscosity syndrome". Readers wishing to pursue the matter further are referred to his book. Another interesting source of information is a short paper by Ehrly [45].

Many pathological conditions affect the rate of coagulation and the firmness of the blood clot and some produce excessive fibrinolysis (e.g. prostate cancer and some hepatic diseases). All this is well shown in the "Atlas de Thrombodynamographie" by Marchal et al. [46]. (These authors prefer this term to the more usual "elasto-" but this arises from a misunderstanding of the technical uses of the term "élasticité": there is, alas, no other word for "springiness" in French.) Unfortunately, so far as is known, all coagulation and "softening" curves follow the same equation (Equn. VI-1): all that change are the values of the three parameters representing rate of coagulation, limiting (theoretical) and actual maxima; also in softening, the rate as defined by the exponential equation. Even so, the "Atlas" is extremely interesting. It should be noted, however, that some haematologists maintain that the thrombelastograph is not sufficiently sensitive and that the charts show characteristic features only when the condition of the patient is obvious from other sources of information.

It is a great pity (or so it seems to the author) that quite a high proportion of the probable readers of this book are not able to read French! It is to be hoped that the younger generation will not have this disadvantage. To translate a whole book, such as that of Larcan and Stoltz into English would be a big task; yet this book, written more from the medical angle than most of the other works quoted, is very valuable. There is a wide bibliography and the rheological concomitance of very many diseases is discussed. Of special interest is the discussion on "sequestration" and "desequestration". The former is defined as follows (present author's translation): "the immobilization or extreme retardation of a certain volume of blood which remained 'trapped'* at the level of different regions of the body, especially pulmonary and hepatosplanchnic areas which no longer participate in the active circulation. This phenomenon is susceptible to disappearance, in some cases spontaneous, or therapeutic."

Rheology of blood vessels

We must close this rather long chapter (though it really has only "sampled" this vast subject) with a brief account of what is known about the rheology of the blood vessels.

Again we are fortunate in having an admirable review article to guide us up to 1971, by Bergel and Schultz [47]. If we take the simple model of a distensible regular cylinder under pressure, it is clear that the distension will be a function of pressure, radius and wall thickness, as well as of the properties of the material (see also Bergel [48]). "The unsteady motion of a liquid through distensible vessels may be described by sets of equations whose solution under prescribed boundary conditions such as input or output flow or pressure enable the missing parameters to be determined for comparison with experiment." Equations are given which describe the motion of a thin-walled isotropic elastic tube with external longitudinal restraint. The properties of the tissue surrounding the vessel were studied in a classical paper by Womersley [49], whose untimely death interrupted pioneer work in this field.

Bergel and Schultz continue with reference to theoretical predictions regarding pulse wave velocity and give a full summary of the work of many authors on this and on the complex modulus and Poisson ratio** of the vessels. In some of the work, thin-walled elastic tubes are taken as a model, for the sake of simplicity: even so, the treatment is extremely complex. Perhaps the best known of the authors quoted is Taylor [50], only one of whose articles can be listed here. Much of the rest of the review article is concerned with the complex problems of flow in such complicated "tubes" as arteries.

Concerning the aorta specifically, there is an interesting paper by Collins and Hu [51] who studied the dynamic stress–strain relations for fresh aortic tissue in relation to traumatic rupture of the aorta. There is a definite stiffening of the tissue with increasing strain rate: the stress–strain relation is

---

* English "trapped" in original.
** The Poisson ratio is the ratio of the transverse strain to the extension strain.

exponential. Another paper on the properties of large arteries comes from Japan (Azuma et al. [52]).

Cerebral haemorrhage

Cerebral haemorrhage has long been a common cause of death but it is only recently that it has been possible to study changes in the walls of the blood vessels without penetrating the skull. X-ray pictures can show up blood-starved areas*. Alternatively, xenon-133 can be used as a radioactive tracer, detected through its γ-rays.

Strokes can sometimes be treated by removing rough places in the main arteries in the neck. Ultra-sound waves can be bounced off the walls of the arteries and, by means of the Doppler effect, the speed of blood flow can be determined. The latest X-ray techniques do not require the injection of opaque substances.

* See article in London "Times" of June 22, 1972.

REFERENCES

[1]   Lalich, J. L. and Copley, A. L., *Proc. Soc. exp. Biol. Med.*, 51: 32 (1942).
[2]   Scott Blair, G. W. and Burnett, J., *Kolloid Z.*, 168: 98 (1960).
[3]   Scott Blair, G. W., in *Flow Properties of Blood and Other Biological Systems* (Eds. A. L. Copley and G. Stainsby), p. 172 (Pergamon Press, Oxford, 1960).
[4]   Fukada, E. and Kailara, M., *J. Soc. Mater. Sci. Japan*, 17: 304 (1968).
[5]   Copley, A. L., King, R. G. and Scheinthal, M., *Biorheology* 7: 81 (1970).
[6]   Copley, A. L., *Biorheology*, 8: 79 (1971).
[7]   Copley, A. L. and King, R. G., *Thrombos. Res.*, 1: 1 (1972).
[8]   Ferry, J. D., *Physiol. Rev.*, 34: 753 (1954).
[9]   Lorand, L., *Physiol. Rev.*, 34: 742 (1954).
[10] Rozenberg, M. C. and Dintenfass, L., *Aust. J. exp. Biol. Med. Sci.*, 42: 109 (1964).
[11] Rozenberg, M. C. and Dintenfass, L., *Nature*, 211: 525 (1966) and many other papers.
[12] Janes, D. E. and Thomas, H. W., *Nature*, 216: 197 (1967).
[13] Scott Blair, G. W. and Burnett, J., *Biorheology*, 5: 179 (1968).
[14] Scott Blair, G. W. and Matchett, R. H., *Haemostasis*, 1 : 93 (1972).
[15] Copley, A. L., *Science*, 101: 436 (1945).
[16] Budtz-Olsen, O. E., *Clot Retraction* (Blackwell, Oxford, 1951).
[17] Rubenstein, E., *Science*, 138: 1343 (1962).
[18] Hartert, H., in *Hemorheology, Proc. 1st int. Conf.*, Reykjavik, 1966 (Ed. A. L. Copley), p. 335 (Pergamon Press, Oxford, 1968).
[19] Scott Blair, G. W., in *Hemorheology, Proc. 1st int. Conf.*, Reykjavik, 1966 (Ed. A. L. Copley), p. 345 (Pergamon Press, Oxford, 1968).

82

[20] Douglas, A. S., in *Biological Aspects of Occlusive Vascular Disease* (Eds. D. G. Chalmers and G. A. Gresham) (Cambridge Univ. Press, Cambridge, 1964).
[21] Scott Blair, G. W. and Burnett, J., *Biorheology*, 5: 163 (1968).
[22] Ferry, J. D. and Morrison, P. R., *J. Amer. Chem. Soc.*, 69: 400 (1947).
[23] Dees, J. E. and Fox, E. H., *J. Urol.*, 49: 503 (1943).
[24] Kristensen, A., *Acta med. scand.*, 77: 351 (1931).
[25] Rubenstein, E., *Diath. haemorrh.*, 17: 552 (1967).
[26] Macfarlane, R. G. and Tomlinson, A. H., *J. clin. Path.*, 14: 320 (1961).
[27] Macfarlane, R. G., Tomlinson, A. H. and Excell, B. J., in *Flow Properties of Blood and Other Biological Systems* (Eds. A. L. Copley and G. Stainsby), p. 377 (Pergamon Press, Oxford, 1960).
[28] Fåhraeus, R., *Biochem. Z.*, 89: 355 (1918).
[29] Harkness, J., *Lancet*, No. 7302: 280 (Aug. 10, 1963).
[30] Knisely, M. H., Bloch, E. H., Eliot, T. S. and Warner, L., *Science*, 106: 431 (1947).
[31] Knisely, M. H., Bloch, E. H., Brooks, F. and Warner, L., *Amer. J. med. Sci.*, 219: 249 (1950).
[32] Knisely, M. H., *Acta anat.* (Basel), 44, Suppl. 41 (1961).
[33] Fåhraeus, R., *Acta med. scand.*, 161: 151 (1958).
[34] Gelin, L-E., *Acta chir. scand.*, 122: 287 (1961).
[35] Gelin, L-E., Bergentz, S-E., Helander, C-G., Linder, E., Nillion, N. J. and Rudenstam, C-M., in *Hemorheology, Proc. 1st int. Conf.*, Reykjavik, 1966 (Ed. A. L. Copley), p. 721 (Pergamon Press, Oxford, 1968).
[36] Meiselman, H. J. and Merrill, E. W., in *Hemorheology, Proc. 1st int. Conf.*, Reykjavik, 1966 (Ed. A. L. Copley), p. 421 (Pergamon Press, Oxford, 1968).
[37] Brooks, D. E. and Seaman, G. V. F., in *Theoretical and Clinical Hemorheology* (Eds. H. H. Hartert and A. L. Copley), p. 127 (Springer-Verlag, Heidelberg–New York, 1971).
[38] Wells, R., in *Hemorheology, Proc. 1st int. Conf.*, Reykjavik, 1966 (Ed. A. L. Copley), p. 415 (Pergamon Press, Oxford, 1968).
[39] Ehrly, A. M., *Med. Klin.*, 61: 989 (1966).
[40] Proewig, F., *Med. Mschr.*, 12: 732 (1958).
[41] Scott Blair, G. W. and Matchett, R. H., *J. Neurol. Neurosurg. Psychiat.*, 35: 730 (1972).
[42] Sanders, H., Thompson, R. H. S., Wright, H. P. and Zilka, K. J., *J. Neurol. Neurosurg. Psychiat.*, 31: 321 (1968).
[43] Ditzel, J., *Brit. J. Ophthal.*, 51: 793 (1967).
[44] Ditzel, J., *Acta med. scand.* Suppl., 476: 123 (1967).
[45] Ehrly, A. M., *Proc. 6th Conf. Europ. Soc. Microcirculation*, Aalborg (Ed. J. Ditzel), p. 62 (Karger, Basel, 1970).
[46] Marchal, G., Leroux, M. E. and Samama, M., *Atlas de Thrombodynamographie* (Service de Propagande, Edition Information, Paris, 1962).
[47] Bergel, D. H. and Schultz, D. L., in *Progress of Physics and Molecular Biology* (Eds. J. A. V. Butler and D. Noble), Vol. 22, p. 3 (Pergamon Press, Oxford, 1971).
[48] Bergel, D. H., *Lab. Pract.*, 15: 77 (1966).
[49] Womersley, J. R., *Wright Air Devel. Centre Tech. Rep.*, WADE-TR 56 (1967).
[50] Taylor, M. G., in *Pulsatile Blood Flow* (Ed. E. O. Abinger), p. 343 (McGraw-Hill, New York, 1964).
[51] Collins, R., and Hu, W. C. L., *J. Biochem.* (Tokyo), 5: 333 (1972).

[52] Azuma, T., Hasegawa, M. and Matsuda T., in *Proc. 5th int. Congr. Rheol.*, Kyoto (Ed. S. Onogi), p. 129 (Univ. of Tokyo Press, Tokyo, 1970).

BOOKS FOR FURTHER READING
(CHAPTERS V AND VI)

[1] Biggs, R. and Macfarlane, R. G., *Human Blood Coagulation and its Disorders* (Blackwell, Oxford, 3rd edn., 1966).

[2] Dintenfass, L., *Blood Microrheology* (Butterworth, London, 1971).

[3] Larcan, A. and Stoltz, J. F., *Microcirculation et Hémorhéologie* (Masson et Cie, Paris, 1970).

[4] Whitmore, R. L., *Rheology of the Circulation* (Pergamon, Oxford, 1968).

Chapter VII

RHEOLOGY OF CERVICAL MUCUS AND SEMEN: MOTILITY OF SPERM

In women, in most domestic animals and in some other species, there is a small passage which lies between the uterus and the vagina which is known as the "uterine cervix". This cervix contains glands that produce a sticky secretion whose rheological properties are important, both to gynaecologists and to veterinarians, for two reasons.

First, their consistency depends on hormone levels, especially those of oestrogen and progesterone; and secondly, as a result of this, they have characteristic properties at the time of ovulation and in pregnancy. A study of these properties may lead to a simple way of observing when ovulation is taking place (or about to take place) in women and to indicate the presence of "heat" in cows and other animals. The farmer (and certainly the bull) can generally tell when a cow is "on heat" from other symptoms; but, especially under extreme climatic conditions of high or low temperature, these symptoms are often not very obvious.

For women, tests on the presence of certain hormones in the urine give an early and reliable indication of pregnancy; but the hormone levels in the cow are too low for such tests to be applicable and it is usually not possible to tell whether a cow is pregnant until the foetus can actually be felt by the hand per rectum, and this is usually some 40 days after conception*. As we shall see later, a fairly reliable test can be done at 28 days from a study of the rheology of the cervical mucus.

To return to women, part of the efficacy of the contraceptive pills, especially those mainly relying on progesterone, depends on their effect in their increasing the impenetrability of the cervical plug.

As early as 1925, Woodman and Hammond [1] discussed qualitatively the consistency of bovine cervical mucus. They reported that the *quantity* of secretion increased during pregnancy and during the oestrous cycle up to

---

* Very recently, it has been claimed that early pregnancy can be detected frcm the presence of a hormone in the milk and also by reflection of ultrasonic waves.

the time of oestrus. They also noted that, except towards the end of the pregnancy, the secretion becomes very thick and sticky. During the oestrous cycle, the secretion becomes very "runny" and plentiful at oestrus ("heat", which occurs just before ovulation) and passes through a thick and impenetrable condition about half way through the 21-day cycle*. The cycle is remarkably regular under normal conditions; much more so than is the 27–28-day menstrual cycle of women. Even so, many common pathological conditions upset the time of ovulation and this is of importance in planning artificial insemination. Woodman and Hammond also noticed that the addition of 0.05 N NaOH to quite a thick mucus renders it what we should now call "spinable": it could be drawn out into long threads, a condition characteristic of the untreated mucus at the time of oestrus. They also studied the effects of various other simple electrolytes on the mucus.

Although there is some evidence of regular changes in the pH of the mucus during the cycle, the effect of adding NaOH cannot be explained only in terms of an increased alkalinity. Meaker and Glaser [2] had studied the pH of the mucus in both the cervix and the vagina as early as 1929.

The earliest work that the author can trace on the rheological properties of human cervical mucus would appear to have been done in France, by Séguy and Vimeux in 1933. Delfs also published a paper in U.S.A. in 1940. From then on, many papers appeared from all over the world, both on women and animals, and, through lack of communication, there was much overlapping. Not all these papers need be discussed in the present book. A few still earlier papers should, however, be mentioned: Rózsa [3], in Hungary, attempted to diagnose bovine pregnancy from the appearance of the mucus as early as 1935. Chernov [4] claimed that an examination of vaginal smears provided an adequate diagnosis for oestrus in cows and Novoselov [5] measured the viscosity of diluted bovine cervical mucus, using a capillary viscometer. In 1940, Day and Miller [6] published very similar suggestions for diagnosing pregnancy in mares. The mucus is described as "thick, sticky and like honey" in pregnancy.

Some explanation of some of these changes appears to have been given first by an American, Palmer [7], working in France. He ascribes the fibrous appearance at the time of ovulation to the presence of folliculin, and the abrupt disappearance of this property to progesterone. (Strangely enough, it is nowhere stated whether he is concerned with women or with some animal.) In a later paper [8], definitely concerned with women, he develops these ideas further. He discusses the changes in body temperature at ovulation and also

---

* The date in the cycle is generally referred, in cows, to oestrus, since they do not menstruate. In women, the "start" of the cycle generally dates from the onset of menstruation.

shows that, in ovariectomized women, oestradiol benzoate produces a "spinable" secretion, and that progesterone inhibits this. He proposes also a rough method for measuring "spinability". These conclusions were verified by Aberbanel [9], who also noted that the thick mucus, artificially produced by ethenyl testosterone or progesterone, was impenetrable to sperm (see below).

The first major quantitative study of these phenomena would seem to be that of Pommerenke, Viergiver and co-workers, who published a number of papers between 1944 and 1947. An excellent review article [10] by these authors appeared in 1946, unfortunately before the programme was completed. However, much of the later work was chemical rather than rheological, and the rheological findings in general confirm those of the earlier workers already quoted. They were perhaps the first to study the undiluted mucus in a viscometer—a filling capillary type (see Chapter II). The minimum "viscosity" (clearly non-Newtonian) and the maximum quantity were found at the time of ovulation.

Many papers were published on similar lines and it is clearly not necessary to list all of them in a book of this kind. A few of the principal workers in this field must, however, be mentioned. (A review of work before 1952 was written by the present author and includes much of the earlier history [11].) His own work, extending over a period of more than twenty years, will be dealt with later.

Certain individual papers, however, have a special interest. Thus an early paper by Parkes [12] anticipated the importance of the consistency of cervical mucus in relation to contraception and di Paula and Lelio [13] discussed possibilities for discriminating between slight and severe amenorrhoea. This depended partly on the fern-like structures of NaCl crystals to be found in certain dried secretions; a phenomenon to be discussed later.

A good survey of the significance of mucus consistency in relation to cattle breeding (up to 1955) is given in the book by Asdell quoted at the end of this chapter. See also an article by Raps [14] and a long article by Roark and Herman [15]. Mention should be made of the softening effect of the enzyme hyaluronidase. This enzyme, present in semen, certainly softens human cervical mucus but, though to be found in bull semen, it appears to have no effect on bovine secretions (see Kaemmerer and Neumann [16] as well as the author's own experiments).

At the beginning of the Second World War, the author's attention was drawn (by his friend the late Prof. S. J. Folley) to the fact that, in the colder parts of Britain, farmers were often failing to notice the symptoms of oestrus in their cows and that, consequently, time was being wasted in failing to get the cows served. (At that time, there was little artificial insemination: "A.I."). In the light of the early work already quoted in this chapter, it should be possible to design a simple apparatus that would make possible a test on the cervical mucus more indicative of the presence of "heat" than mere subjective observations.

The two most obvious rheological properties to measure were the "flow elasticity" (recoil when mucus is pushed out of a capillary and the pressure is released), and the "spinability"*. The difficulty in measuring the latter was that it seemed that, although "threads" could be formed by drawing out the mucus between the finger and thumb, any sample extended by mechanical means invariably failed because the mucus came away from any form of holder that could be thought of at the time. Many forms of "grip" —rubber, plastic, cork, metal, etc., were tried without success. It was not until much later (1967) that Dr. Weis, from Israel [17], suggested the use of Bassequer eye forceps and these proved highly successful. The length of "thread" that can be drawn without breaking is approximately proportional, over a reasonable range, to the speed of extension. The apparatus is shown in Fig. VII-1.

The author and his colleagues wrote many papers on rheological methods for measuring flow elasticity and consistency of cervical mucus. Most of these were published before 1952 and are summarized in the article already quoted [11]. Only the more important will be discussed here.

The first papers were published in 1941 (Scott Blair et al. [18]). One of these will be quoted because in it, the authors described a very simple little apparatus (perhaps unfortunately called the "oestroscope"**: which should mean "looking at oestrus"). This is shown in Fig. VII-2 and consists of a glass tube (A), about 2 mm bore and 11 cm long, attached to a hypodermic syringe. The side-tube (B) is closed with the thumb and mucus is sucked into the capillary. (If the cow is anywhere near oestrus, the mucus is so plentiful that it can be collected from the end of the vagina, or even the vulva.) The mucus is then gently extruded from the capillary, not far enough to allow it to fall away from the open end, the thumb is released from B, and the recoil is recorded.

This test was never widely used in practice. Farmers in most countries are "traditionalists" and do not take readily to such tests. Some use was made of it, however, in Israel, where Dr. Braun (unpublished thesis and ref. 19)

---

* Until recently, rheologists almost always referred to this property by its German name: "Spinnbarkeit".

** Oestroscopes were manufactured by Messrs. Arnold and Sons, 54 Wigmore Street, London, W. 1., England.

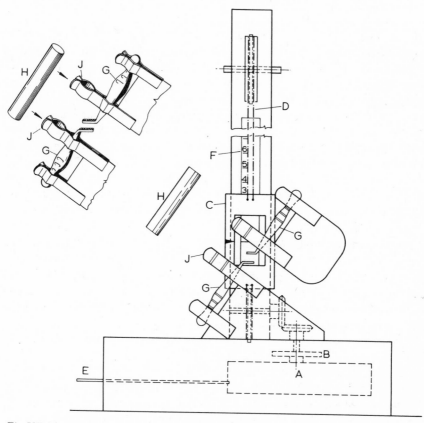

Fig. VII (1). Diagram of apparatus for measuring "spinability" of mucus. Reproduced from: Burnett, J., Glover, F.A. and Scott Blair, G.W., Biorheology, 4(2): 41 (1967).

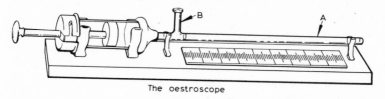

The oestroscope

Fig. VII (2). The oestroscope. Reproduced from: The Oestroscope by Scott Blair, G.W. et al. Manufactured by Arnold and Sons, London.

was able to use discrete "blobs" (indexes) of mucus instead of extruding a column from a tube. The present author never succeeded with this method. Another Israeli worker, J. Crane, was tragically killed in a road accident while experimenting with the test and the later pregnancy tests. Some of his results were published posthumously [20]. Crane used the "index" method proposed by Braun and determined an elastic modulus and a viscosity for the mucus. He concluded that the product of these gave a sharper indication of the time of oestrus than did the elastic recoil alone. These experiments were done following a large number of cycles in a single cow, though a few other cows were also tested.

Several large-scale tests were made elsewhere. Herman and Horton [21], in 1948 in the U.S.A., tested 80 cows with considerable success, and Blackburn and Castle [22], in Scotland, made a careful statistical study of the method (over 750 samples) and found that oestrus occurred, except in certain pathological conditions, within a day or so of the maximum elastic recoil in 94% of cases. Nevertheless they doubted whether the test would find much practical application. Much later, Salama et al. [23], in Egypt, showed that the test could be effectively used for dairy buffaloes. This work was described more fully in a doctorate thesis, Warsaw Agricultural University. (As a little light relief, the author tried to get this test used at a French Zoological Garden, but the veterinarian reported that "les bufflesses étaient trop méchantes".)

Tests for bovine pregnancy

Since, as already stated, mucus from pregnant cows has a high consistency, it seemed worth while to measure this consistency and to find out empirically whether an arbitrary value could be found so that early pregnancy could be distinguished from the dioestrous condition (halfway through the cycle) in a non-pregnant animal, when the consistency is also rather high. This technique was developed gradually and many papers were published over some twenty years. Finally, an apparatus was made which measured the rate of flow of the mucus out of an emptying capillary under a constant arbitrarily selected air pressure. It was found that the rate of flow was proportional to the square root of the pressure, so that repeats could be made at different pressure and the results directly compared. The apparatus (also manufactured by Messrs. Arnold and Sons) is shown in Fig. VII-3. (The lettering refers to the different taps used when applying and releasing the air pressure from the small pump; see Scott Blair and Glover [24].) Only the last series of experiments will be summarized here.

A few pathological cases were discarded: infected vagina with purulent

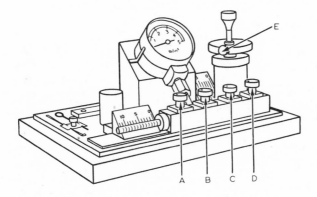

Fig. VII (3). Consistometer for detecting pregnancy in cows. Reproduced from: The Consistometer, Scott Blair, G.W. Manufactured by Arnold and Sons, London.

mucus, ovarian cysts mainly. The most effective time for detecting pregnancy seemed to be 28 days after conception. This is, of course, not nearly as early as that available for women using hormone tests, but is well in advance of the (approximately) 40 days required for the rectal method. 351 tests were made on three breeds of cow, Shorthorn, Friesian (Holstein) and Guernsey. Slightly better results were obtained if a small difference was made in the critical level of consistency indicative of pregnancy, for the three breeds. Among cows later confirmed as pregnant, 92% were correctly diagnosed at 28 days. Among cows which had, in fact, not been served, 95% were correct, but for non-pregnant cows which had been served, only 75% were correctly diagnosed. The reason for this last low figure was easily found. When after some 40 days, the veterinary examination showed no pregnancy, the rheological test was repeated and also showed "not pregnant". In other words, as already suspected by the veterinarians, there was a considerable incidence of early foetal deaths.

Blackburn and Castle [25] carried out a similar study on quite a large number of cows, but did not discriminate between the lengths of time after service. Probably as a result of this, their success level was not as high as Scott Blair and Glover's, though in general they confirmed the earlier results. They foretold, quite rightly, that the method would probably not be used widely in practice partly because it requires considerable skill to take the sample, scanty in pregnancy, from the cervical os, and also because, with frequent early foetal deaths, the practical farmer would prefer to know that his cow was really likely to produce a calf after waiting 40 days, rather than to be told that she was pregnant at 28 days, but might quite possibly not come to term.

Before passing from bovine to human cervical mucus, brief mention should be made of two other phenomena which, although not rheological, are closely linked with the rheological changes. In the U.S.S.R., Aïzinbudas [26] and his colleagues found changes in the electrical resistance within the cervix during the sexual cycle and in pregnancy. Babucheva [27] confirmed this on 700 cows. It is uncertain who first noticed the curious fern-like structures of NaCl crystals which may be seen under a low-power microscope in certain dried-out secretions. They are also sometimes described as "palm-leaf" or "PL" structures. In 1948, Rydberg [28] observed that the NaCl content of mucus is high in the ovulatory phase; but the earliest full account would seem to be that given in Zondek's book [29].

Zondek pointed out that "PL" is not confined to cervical mucus, but is found in some other body fluids. In cervical mucus, however, it is certainly most marked at about the time of ovulation, and absent in pregnancy. In fact, Professor Zondek suggested to the author that absence of "PL" might prove to be an adequate indication of pregnancy in the cow, but this is, unfortunately, not a reliable test.

In the work described in Ref. 24 the authors examined all the dried-out secretions under the microscope, but the diagnosis of pregnancy at 28 days was not as good as that given by the rheological test: only 90% of pregnant cows, 78% of cows that had not been served and 62% of cows that had been served but were not pregnant.

Pozo Lora et al. [30] (writing in Spanish) classified the crystals in cervical mucus in a variety of different types. Earlier workers, notably Berman [31] and de la Fuente and Gálvez [32], had noted characteristic differences in the types of crystals; but it was not until 1956 that Fedrigo and Bergamini [33] suggested that the appearance of the crystals and their disaggregation depends on the concentrations of mucinase and trypsin. Some of the patterns described by Pozo Lora are shown in Fig. VII-4.

Fedrigo and Bergamini measured the viscosity of diluted secretions (from cows) in an Ostwald capillary viscometer, making a special study of the action of various enzymes.

Human secretions

In a series of papers, too numerous to list, Zondek, in 1954, made a further study of human secretions. He found that ovulation mucus loses its "PL" on mixing with semen. Pregnant secretions, to which NaCl has been added,

Fig. VII (4). Crystal patterns of dried cervical mucus ("PL"). Reproduced from: Pozo Lora, R., et al., Archivos de Zootecnia, 1958, Imprenta Moderna, Cordoba, Spain.

will develop "PL" when dried. But pregnant secretions, even in the presence of oestrogen, will not normally induce "PL" and this fact might form the basis for a pregnancy test (though, surely, not nearly as sensitive a test as the hormone tests already available for women, including the Ascheim–Zondek!). Traces of "PL" in mucus during pregnancy indicates a placental insufficiency and may foretell an abortion. Zondek did not find that the "PL" test was sensitive enough to indicate the precise date of ovulation.

Although "PL" certainly consists of crystals of NaCl, Scott Blair and Glover [34] showed that the degree of "PL" formation is not a simple function of either moisture content or NaCl concentration.

Following the earlier work described above, gynaecologists were not slow to see the possibilities inherent in the use of rheological tests to determine the time of ovulation (in spite of Zondek's somewhat pessimistic conclusion); and also it might well be that the rheological properties of the mucus could be related to certain pathological conditions.

The present author worked for some years, with Mr. A. F. Clift*, a gynaecologist, in both ante-natal and post-natal clinics. Although the sample of mucus must be taken with great care from the os of the cervix (forceps are used which do not even touch the tissue of the cervix), this procedure is much simpler and less trying to the patient than is an endometrial biopsy, and basal temperature tests may be vitiated by extraneous factors.

In a preliminary paper, Clift et al. [35] measured the varying pressure required to extrude the mucus from a capillary at a constant rate of extrusion. As an "index" of consistency, they took the logarithm of the slope of the flow curve. The results are clearly shown in Fig. VII-5. The time scale is in days (of menstrual cycle) for non-pregnant women and in weeks for pregnant women. Samples were taken from 300 women. There was a marked minimum at the time of ovulation (as confirmed by endometrial biopsy and temperature) and a less marked minimum before menstruation. (Naturally, samples could not be taken during menstruation.) In a very few cases, there was an over-lapping of consistency between samples from women in early pregnancy and those at dioestrus, when the (non-pregnant) consistency was at its maximum. Some patients showing relatively low consistency in early pregnancy, later aborted.

In an earlier paper, Clift [36] described a modified form of the oestroscope, suitable for tests on women, which he called the "menstroscope". He also stressed the significance of "spinability" (called, at that time, by its German

---

* For non-British readers: in Britain, surgeons, however well-qualified in medicine, prefer to be called "Mr." and not "Dr."

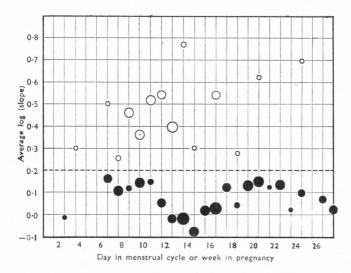

Fig. VII (5). Average log (slope) of cervical secretions at different stages of menstrual cycle or of pregnancy in 86 non-pregnant women (black circles) and 35 pregnant women (open circles). Areas of circles are proportional to the number of cases included in the average, the smallest circle representing one case. Reproduced from: Clift, A.F. et al., Lancet, 1154 (1950).

name, "Spinnbarkeit") and tack, or tackiness. He also suggested that the addition of donated cervical mucus of suitable consistency might be added to semen for artificial insemination in cases of "cervical hostility". (This last phenomenon, i.e. the incompatibility of an individual semen with an individual cervical mucus, has been alleged to account for some cases of infertility.)

Clift and his colleague Hart [37] adopted for use with women the "index" method already in discussion in relation to bovine secretion. The data on this occasion did not show any evidence of a "menstrual minimum" such as was found for the overall consistency, using the continuous column method, but no readings were taken between the 28th and 40th days of the cycle.

Thirty-seven pregnant women all showed a higher consistency of mucus than did any of the non-pregnant women, except in certain pathological cases. Later, Hart [38] described a rising sphere viscometer suitable for materials such as cervical secretions.

Clift's work was followed up in a number of Gynaecology Departments in British hospitals. Thus Shotton [39] confirmed many of Clift's findings and studied the effects on the mucus of various hormonal abnormalities; and Harvey et al. [40] found, statistically, an inverse relation between the consistency using Scott Blair's method, and the flow elasticity. They concluded that basal temperature gave a more reliable indication of ovulation than did the elastic recoil.

Quite independently, however, Beller and Vogel [41], in Germany, concluded that the rheological diagnosis was more reliable than the basal temperature test (see also a review article by the same authors [42]). (Beller published a number of other papers on much the same lines between 1960 and 1962.) The work of Hwang et al., covering many types of mucus, will be discussed in the next chapter.

Penetration of cervical mucus by spermatozoa

In an early paper, von Khreningen-Guggenberger [43] showed that sperm cannot move vertically upwards in a salt solution, but they do so in vaginal mucus. This suggested to him that substances might be found that would reduce the consistency of cervical mucus and so prove an effective method of birth control. The modern view is exactly the contrary!

Lamar et al. [44] placed little "blobs" of mucus and semen in a capillary, separated by air bubbles. The sperm thus had to pass along the very thin, moist layer on the wall of the tube. "Viscosities" of mucus were roughly measured but showed quite irregular curves during the sexual cycle. But the penetration by the sperm proved to be maximal at about the 14th day of the cycle, as did also the pH and absence of leucocytes. Barton and Wiesner [45] showed that cervical mucus and semen do not mix: the individual sperm move across the boundary between these two phases. In some cases of sterility, it seems that the sperm are inhibited from passing through the mucus.

Working on 80 cows, Herman and Horton [46] showed that the penetrability of sperm is maximal 6–12 hours after the start of oestrus, as the volume of secretion tends to decrease, and the number of leucocytes increases during oestrus. There is also a rise in vaginal temperature.

Harvey, in a series of articles of which only one is listed here [47], studied the penetrability of sperm into human cervical mucus, arranged on a special glass slide to avoid problems of surface tension. The distances of penetration in a given time were measured and it was concluded that the properties of the mucus were of more significance than those of the spermatozoa.

The flow of sperm in semen

The way in which spermatozoa swim in semen or in Newtonian liquids is primarily a hydrodynamic rather than a rheological problem. Nevertheless, some brief account of work in this field should be included in this chapter. The principal worker in this field was Lord Rothschild, following studies by Sir Geoffrey Taylor on the hydrodynamics of the swimming methods of eels. Only one of Rothschild's papers need be quoted here [48]. Summarizing, sperm tend to swim in shoals. With a parabolic velocity gradient, dead sperm tend to face up-stream in the lower half and down-stream in the upper half of the stream. This is caused by the higher specific gravity of the head as compared with the tail. Live sperm face up-stream in both halves. (This behaviour is called "positive rheotaxis".) When the direction of the stream is reversed, the dead sperm re-orientate, as might be expected hydrodynamically. Live sperm behave in a more complicated and unexplained way.

The "wave motion" of sperm was, of course, known long before 1961. (An excellent summary of early work, going back as far as 1792, is given by Walton [49].)

Walton also studied the orientation of optical and electrical methods. The latter method, a measure of impedance, was developed by Rothschild in 1948–1949 but the present author got the impression, many years later, that Lord Rothschild was no longer enthusiastic about this technique. Walton showed that the time required for a suspension of elongated cells, orientated hydrodynamically by shearing, to regain their (then supposedly) random distribution was a measure of their motility, an important factor in relation to the fertility of the semen.

Following the later findings of Rothschild, the present author suggested to his colleague, Mr. F. A. Glover, in 1966, that it would be interesting to try placing bull semen between two disks, the upper disk being transparent, rotating the disks relative to one another so as to align the sperm hydrodynamically and then, on stopping the rotation, to measure the time taken for the sperm to regain their "biological" orientation (see Glover [50]).

Glover and Scott Blair [51] had also found that the viscosity of bull semen (using a Couette-type viscometer) first rose, then fell and then finally rose again when the temperature was raised as quickly as possible from 20 to 40°C. (The semen was cooled to 20°C when taken from the bull because the seminal fluid starts to damage the sperm as soon as they are mixed and this effect is minimal at 20°C. The sample was tested within a few minutes after collection.) As shown in Fig. VII-6, the effect is most marked at a relatively low shear rate. It was suggested that the initial rise in viscosity was caused

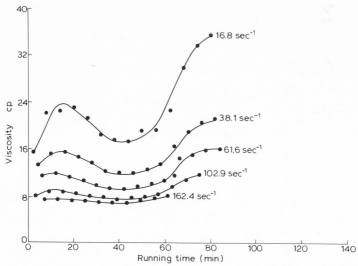

Fig. VII (6). Changes in viscosity of bull semen at various shear rates and temperatures. Reproduced from: Glover, F.A. and Scott Blair, G.W., Biorheology, 3: 189 (1966).

by an activation of the sperm as a result of the rise in temperature, followed by a fall resulting from their death and a further rise produced by post-mortem clumping.

There were, therefore, two independent possible methods of assessing sperm motility. Their value would have been greatly increased had it been possible to establish direct correlations between them. Unfortunately this was not done: a correlation with the "effectiveness" of semen from individual bulls as determined by the proportion of "successful" inseminations was insignificant; but this is hardly surprising in view of the large number of other factors involved. (Later work, however, showed a fair correlation between subjectively assessed motility and the data from the disk shearing test.)

Rikmanspoel [52] has studied the tail movements of bull spermatozoa, which have a fibrous sheath. Active contractile elements along the tail maintain the amplitude of the wave. Rikmanspoel [53] has also studied the elastic properties of sea-urchin sperm flagellum, applying the theory of the vibration of thin rods. If sufficiently flexible, the flexibility depends on the wavelength of the travelling wave (if the "rod" is sufficiently flexible, i.e. there is no fibrous sheath). The stiffness is produced mainly by longitudinal fibres with a Young's modulus of about 10^8 dynes/cm^2, which develop a tension of about 1.6×10^8 dynes/cm^2, comparable to that found in muscle fibres.

A good review article on sperm flagella movements was published by Holwill [54].

The consistency of semen

Very little work appears to have been done on the "viscosity" (certainly anomalous) of semen, even in man. With farm animals, the situation is extremely complex; for example, the boar produces enormous quantities of semen whose consistency changes greatly during a lengthy process of ejaculation; whereas the bull, a much larger animal, produces, at each ejaculation, only a small quantity of relatively liquid semen. On the other hand, in the latter case, the supply is very quickly renewed and a bull left free with a cow "in season" will mount her very frequently in the course of a day.

Using bull semen, Szumowski [55] claimed that "viscosity" and sedimentation rates of sperm were related to fecundity. The mean relative viscosity was 3.74. The time during which motility can be maintained had little to do with the initial viscosity. Motility and viscosity were correlated. Semens with high sperm counts retained their fertility longer than did those with fewer sperm: but all these conclusions appear to depend on semen taken from a single bull.

Additional notes

1. It appears that (bovine) cervical mucus can be used as a substitute for respiratory mucus in the frog! (Eliezer [56]).

2. As mentioned above, one practical aspect of the importance of the consistency of cervical mucus lies in its effect in allowing or preventing the passage of sperm through the cervical canal. At the time of writing (August 1972) the relative merits of different hormones as contraceptives are still very much in dispute. The slightly increased danger of thrombosis and perhaps other conditions, following the use of oestrogens must be balanced against the apparently slightly lower efficiency of progesterone and the side-effects that it sometimes produces. For a very thorough test of the effects of Norgestrel (a low-dose progesterone-only contraceptive pill), see Eckstein et al. [57].

REFERENCES

[1] Woodman, H. E. and Hammond J., *J. agric. Sci.*, 15: 107 (1925).
[2] Meaker, S. R. and Glaser, W., *Surg. Gynec. Obstet.*, 48: 73 (1929).

100

[3] Rózsa, L., *Közlemének az Összehasonlító, életés kértan köréböl.*, 26: 407 (1935) (In Hungarian).
[4] Chernov, V. I., *Probl. Zhivotnostva*, 5: 102 (1933).
[5] Novoselov, S. I., *Sborn. Rabot Leningrad Vet. Inst.*, p. 92 (1935).
[6] Day, F. T. and Miller, W. C., *Vet. Rec.*, 52: 711 (1940).
[7] Palmer, R., *C. R. Soc. Biol.* (Paris), 135: 366 (1941).
[8] Palmer, R., *Ann. Endocr.* (Paris), 5: 137 (1944).
[9] Aberbanel, A. R., *Endocrinology*, 39: 65 (1946).
[10] Pommerenke, W. T. and Viergiver, E., in *Problems of Fertility* (Ed. E. T. Engle) (Princeton Univ. Press, Princeton, N. J., 1946).
[11] Scott Blair, G. W., in *Deformation and Flow of Biological Systems* (Ed. A. Frey-Wyssling), p. 447 (North-Holland Publ. Co., Amsterdam, 1952).
[12] Parkes, A. S., *Proc. Soc. Study Fertil.*, No. 5: 20 (1953).
[13] di Paula, G. and Lelio, M., *J. clin. Endocr. Metab.*, 13: 974 (1954).
[14] Raps, G. R., *J. Amer. vet. med. Ass.*, 114: 206 (1949).
[15] Roark, D. N. and Herman, H. A., *Univ. Mo. agric. Res. Sta. Res. Bull.*, p. 445 (1950).
[16] Kaemmerer, K. and Neumann, H. G., *Z. Tierzücht Züchtungs-biol.*, 58: 416 (1950).
[17] Burnett, J., Glover, F. A. and Scott Blair, G. W., *Biorheology*, 4: 41 (1967).
[18] Scott Blair, G. W., Folley, S. J., Malpress, F. H. and Coppen, F. M. V., *Vet. Rec.*, 53: 693 (1941).
[19] Braun, I., *Proc. 2nd int. Congr. Rheol.*, Oxford, 1953 (Ed. V. G. W. Harrison), p. 324 (Butterworth, London, 1954).
[20] Crane, J., Reiner, M. and Scott Blair, G. W., *Bull. Res. Coun. Israel E*, 8: 87 (1960).
[21] Herman, H. A. and Horton, O. H., *J. Dairy Sci.*, 31: 679 (1948).
[22] Blackburn, P. S. and Castle, M. E., *Brit. vet. J.*, 115: 399 (1959).
[23] Salama, A., Shalash, M. R. and Hoppe, R., *Biorheology*, 4: 127 (1967).
[24] Scott Blair, G. W., and Glover, F. A., *Brit. vet. J.*, 113: 417 (1957).
[25] Blackburn, P. S. and Castle, M. E., *Brit. vet. J.*, 118: 321 (1962).
[26] Aızinbudas, L. B. and Dovil'tis, P. P., *Zhivotnovodstvo*, 24: 68 (1962) and 28: 84 (1966).
[27] Babucheva, L.Ya., *Zhivotnovostvo*, 27: 74 (1965).
[28] Rydberg, E., *Acta obstet. gynec. scand.*, 28: 172 (1948).
[29] Zondek, B., *Recent Progress in Hormone Research* (Academic Press, New York, 1953).
[30] Pozo Lora, R., Scott Blair, G. W., Ayalon, N. and Glover, F. A., *Consejo Superior de Investigaciones Cientificas*, Dept. Zootecnia, Córdoba, 1958 (in Spanish).
[31] Bergman, P., *Acta obstet. gynec. scand.*, 29: 296 (1950) and *Fertility Sterility*, 4: 183 (1953).
[32] de la Fuente, F. and Gálvez, J., *Acta ginec.* (Madr.), 5: 149 (1954) and 5: 352 (1954).
[33] Fedrigo, G. and Bergamini, L., *Nuovo Vet.*, 32: 7 (1956) and 32: 37 (1956).
[34] Scott Blair, G. W. and Glover, F. A., *Nature*, 179: 420 (1957).
[35] Clift, A. F., Glover, F. A. and Scott Blair, G. W., *Lancet*, p. 1154 (June 24, 1950).
[36] Clift, A. F., *Proc. roy. Soc. Med.*, 39: 1 (1945).
[37] Clift, A. F. and Hart, J., *J. Physiol.* (Lond.), 122: 358 (1953).
[38] Hart, J., *J. sci. Instrum.*, 31: 182 (1954).
[39] Shotton, D. M., *Proc. Soc. Study Fertil.*, 3: 9 (1951).
[40] Harvey, C., Linn, R. A. and Jackson, M. H., *J. Reprod. Fertil.*, 1: 157 (1960).
[41] Beller, F. K. and Vogel, H., *Z. Geburts. Gynäk.*, 158: 58 (1962).

[42] Beller, F. K. and Vogel, H., in *Flow Properties of Blood and Other Biological Systems* (Eds. A. L. Copley and G. Stainsby), p. 248 (Pergamon Press, Oxford, 1960).
[43] von Khreningen-Guggenberger, *J. Arch. Gynäk.*, 153: 64 (1933).
[44] Lamar, J. K., Shettles, L. B. and Delfs, E., *Amer. J. Physiol.*, 129: 234 (1940).
[45] Barton, M. and Wiesner, B. P., *Brit. med. J.*, p. 606 (Oct. 26, 1946).
[46] Herman, H. A. and Horton, O. H., *J. Diary Sci.*, 31: 679 (1958).
[47] Harvey, C., *J. Obstet. Gynaec. Brit. Emp.*, 61: 480 (1954).
[48] Bretherton, F. P. and Rothschild (Lord), *Proc. roy. Soc. B*, 153: 490 (1961).
[49] Walton, A., *J. exp. Biol.*, 29: 520 (1952).
[50] Glover, F. A., *Nature*, 219: 1263 (1968).
[51] Glover, F. A. and Scott Blair, G. W., *Biorheology*, 3: 189 (1966).
[52] Rikmanspoel, R., *Biophys. J.*, 5: 365 (1965).
[53] Rikmanspoel, R., *Biophys. J.*, 6: 471 (1966).
[54] Holwill, M. E. J., *Physiol. Rev.*, 46: 696 (1966).
[55] Szumowski, P., *Rec. Méd. vét.*, 124: 124 (1948).
[56] Eliezer, N., *Proc. 1st int. Congr. Biorheol.*, Lyon., 1972, published in *Biorheology*, 10: (1973).
[57] Eckstein, P., Whitby, M., Fotherby, K., Butter, C., Mukherjee, T. K., Burnett, J. B. C., Richards, D. J. and Whitehead, T. P., *Brit. med. J.*, 3: 195 (1972).

BOOKS FOR FURTHER READING

Asdell, S. A., *Fertility and Sterility* (J. and A. Churchill, London, 1955).
Melrose, D. R., in *The Semen of Animals and Artificial Insemination* (Ed. J. P. Maule), Commonwealth Bur. Anim. Breed. Gen. Commun., No. 15: 31 (1962).

Chapter VIII

RHEOLOGY OF SYNOVIAL FLUIDS:
FRICTION AND LUBRICATION IN JOINTS

General rheology

After coronary disease and cancer, perhaps the most serious problems in medicine, especially in Great Britain, are concerned with the prevalence of rheumatism and similar conditions. Bronchial diseases are also very common and will be discussed in the next chapter.

Synovial fluid

It is perhaps rather surprising that attention was not paid much before the 1930's to the rheology of the "fluid" that lubricates the joints. (We write "fluid" without implying that this material is a true fluid in the rheological sense of the term.) It will perhaps be best to follow the progress of the work more or less in historical sequence. As elsewhere in this book, the selection of material is bound to be somewhat arbitrary, since in recent times, far too many papers have been published to quote all of them.

As early as 1925, Schneider [1] studied the flow of synovial fluids, using a microcapillary viscometer and attempted to relate the viscosities to pathological conditions. He claimed that the viscosities were lower than normal in most chronic illnesses with high values following sudden death. But the range of relative viscosities was enormous: 4–1490 for corpses, and 500–1500 for normal healthy people.

Among early workers, Panizza [2] should also be mentioned. This author used an emptying capillary viscometer, and showed that the synovial fluid from different joints (of cattle) had characteristically different viscosities, but he had little to say about viscous anomalies. The differences were ascribed to differences in the mucopolysaccharides. A similar study was made by Bywaters [3] a few years later. Panizza also summarized still earlier work, now perhaps of little interest to modern readers.

It was not long before the importance of hyaluronic acid in synovial fluids came to be realized. The enzyme hyaluronidase softens the mucus, as it does also human (and monkey) cervical mucus (see Chapter VII). Chain and Duthie [4] showed that it also increases the permeability of the dermal layer of skin.

Davies [5] studied the volume, viscosity and nitrogen content of synovial fluids, using an emptying capillary. He was one of the first to consider seriously the general problems of joint lubrication, finding characteristic differences between the friction in different joints (cattle). He gave a good bibliography of earlier work, and quoted previous findings by various authors relating viscosity to the quantity of mucopolysaccharide, but did not discuss viscous anomalies.

Blix and Snellman [6] were among the first to measure the intrinsic viscosity of mucus from various sources, including the joints. The "intrinsic viscosity", written as $[\eta]$, has also been called by a number of other names; but, although there is nothing particularly "intrinsic" about it and it is not really a viscosity, the names proposed later appear to be no better. It depends on an equation proposed by Huggins, which, for many materials, relates relative viscosity to concentration.

First, the nomenclature: η_{rel}^{-1} is called "specific viscosity" (η_{sp}) and this, divided by concentration (C) is called "reduced viscosity" (η_{red}). It is found that, if η_{red} is plotted against concentration, straight lines are often obtained which extrapolate to an intercept ($[\eta]$) on the η_{red} axis. This corresponds to the equation $\eta_{red} = [\eta] + k [\eta] C$ of Huggins and k is called the "Huggins constant". (For more concentrated solutions, higher power terms must be added.)

Blix and Snellman also calculated molecular dimensions from these data and from measurements of flow birefringence. Ragan [7] found difficulties in getting enough synovial fluid from healthy living subjects, so he measured viscosities of solutions in NaCl and also took samples from corpses. He found that, in general, the viscosity of the mucus is lower in pathological than in normal conditions.

One of the most important early papers is that of Ropes et al. [8]. In spite of known non-Newtonian behaviour, they used an Ostwald capillary viscometer, working with bovine mucus. They studied the effects of changes in pH and in salt concentrations. The relative viscosity passed through a rather "flat" maximum at pH values from about 5 to 7.

They also devised an "oiliness tester". "Oiliness", sometimes unfortunately called "lubricity", is a complex property which depends by no means entirely on viscosity and is important in the lubricant industry.

Synovial fluid does not appear to be, technically, a particularly good lubricant, but it has many functions other than that of merely reducing friction. Under most pathological conditions, the body reacts by producing more synovial fluid, but not enough more hyaluronic acid to maintain the viscosity.

Ropes et al. give average viscosities for a number of different pathological conditions, but the number of patients was small and the range of viscosity large. As many later workers found, viscosity cannot be regarded as specifically diagnostic.

But the authors made other interesting suggestions, including the idea that a fall in hyaluronic acid content may account for some of the degenerative changes in old age. Even at this early date (1947), 111 references are given in this paper. Two years later, Rimmington [9] wrote an excellent review article on the work published at that time. Between the years 1950 and 1953, Ogston and Stanier [10, 11] made intensive studies of hyaluronic acid and synovial fluids. Refractive indices were measured, and viscosity, using both an Ostwald and a (modified) Couette viscometer. The logarithm of the relative viscosity was found to be roughly linear with the glucosamine concentration. The intrinsic viscosities were also measured and, using a method proposed by Simha, an "axial ratio" (length to width) of about 590 was calculated. A loose, sponge-like structure of rod-like particles was suggested.

Hyaluronic acid is a protein with a molecular weight of about 10^6 and consists of rod-shaped molecules about 2000×33 Å in size.

But very different results were got from flow birefringence and sedimentation rate experiments. The long-chain molecules of hyaluronic acid appear to be coiled up, enveloping much of the solvent. The degree of degradation of hyaluronic acid was also studied.

In another paper, Ogston et al. [12] find shear elasticity in synovial fluids, using the Couette-type viscometer.

The year 1953 marks a convenient, if arbitrary date to separate the older experiments from the more modern work.

Sundblad [13] has written a long article, reviewing work up to that date with about 100 references. Sundblad's own experiments were extensive, and he also gave a very full account of what was known at the time about hyaluronic acid, the substance that is mainly responsible for the viscous properties of synovial fluid. The enzyme hyaluronidase, that breaks down hyaluronic acid, was originally obtained by Chain and Duthie [4] from mammalian testes, but it is to be found in many sources, associated with connective tissue. It was suggested that a study of the mucus might well lead to useful information about the state of the connective tissue surrounding the joints. The principal aim of Sundblad's experimental programme was "to supply information on the degree of polymerization of the synovial hyaluronic acid in normal and pathological fluids by means of determinations of viscosity and hyaluronic acid concentration".

For routine measurements, an ordinary Ostwald viscometer was used, but for small samples, a "semi-micro type" was required. For studying anomalous viscosity, a horizontal capillary was used, sealed at each end to wider graduated tubes in which the movement of a meniscus could be followed at a series of constant pressures. The rate of shear (at the wall, presumably) was calculated as $\dfrac{8\,V}{3\,\pi\,R^3}$ (using the symbols to be found in this book; V is volume per second, R is radius of capillary); but it is not clear why "8/3" was used instead of the "4" derived directly from the Poiseuille equation. The wall stress is given as usual by $P\,R/2\,L$, where P is the pressure and L is the length of the tube.

Graphs are shown giving a curvilinear relation between relative viscosity and concentration (not surprisingly!) for solutions containing hyaluronic acid in various degrees of polymerization. Of more interest are the straight lines obtained plotting η_{red} against η_{sp} (it is usual to plot against concentration) both for synovial fluids and for certain hyaluronate preparations in phosphate NaCl. But it must be admitted that there are very few points on the straight lines.

The author gives tables of ranges and averages for intrinsic viscosities for a number of pathological conditions, but the range of viscosities is very wide and the number of patients tested is small. However, it appears to be significant that the intrinsic viscosity does tend to fall with ageing in all the conditions listed, except degenerative joint disease, where there is a considerable rise comparing age groups 27–52 with 53–82.

Even the summary of this long paper cannot be dealt with here in detail. The author concludes that the hyaluronic acid concentration and the relative viscosity were of only limited value in differentiating pathological conditions but he believed that the intrinsic viscosity had some diagnostic value. Changes in viscosity produced by incubating the mucus at 37°C were also claimed to be helpful.

Scott Blair et al. [14] examined over 60 samples of human synovial effusions in a capillary viscometer. There was a wide variation of viscosity and only the more viscous samples showed appreciable non-Newtonian behaviour. A rather strange and unexplained equation was found to hold remarkably well. This postulates a parameter having the dimensions of a viscosity, equal increments of which correspond to progressive halvings of the remaining stress: i.e. the viscosity is linear with the logarithm of the applied pressure.

The viscometer consisted of a horizontal capillary attached at each end to wider tubes bent into a vertical position and the head vs. time curve was determined for small increments of head (Δh) and time (Δt). Fig. VIII-1 shows good straight lines for seven samples of widely differing viscosity, plotting $\Delta t/\Delta$ (log h) vs. log h. No appreciable "sigma effects" were found, but there was an anomalous "length" effect: a slight increase in viscosity with increasing capillary length, i.e. a shear thickening. Shearing and stirring produced a slight increase in viscosity. No correlation was found, with the rather crude apparatus available at the time, with ESR (erythrocyte sedimentation rate), sex, age, or duration and nature of the disease. A further complication is that the removal of the effusion from a joint itself tends to lower the viscosity of the mucus.

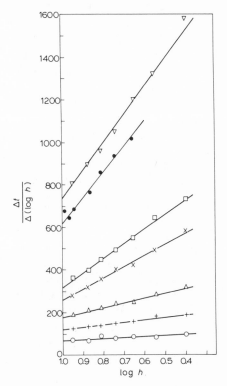

Fig. VIII (1). Viscosity of synovial effusions (differential definition) plotted against logarithm of stress (t in sec, h in cm). The seven types of points refer to seven different samples. Reproduced from: Scott Blair, G.W. et al., Biochem. J., 56: 504 (1954).

In the same year, Jensen and Koefoed [15] studied the behaviour of solutions of pure potassium hyaluronate, using the Scott Blair oestroscope (see Chapter VII). Elastic recoil and "viscosity" (overall consistency) were measured at different concentrations. The maximal recoil was found at a concentration of 0.25%. Crude hyaluronate (from the umbilical cord) and bovine synovial fluid were also tested. A theory of elasticity for hyaluronic acid, based on rubber elasticity theory, was developed. It is assumed that the hyaluronate ions adhere to the glass wall and form a network throughout the body of the material. Flow is associated with the breaking of hydrogen bonds. Such a structure is eventually destroyed by thermal rearrangement. Comparing the author's results with those of Scott Blair on cervical mucus, they postulate viscous flow of solvent through the gaps in the hyaluronate so that the recoil is not independent of the length of the displacement. The same authors [16] suggest that flow potentials of potassium hyaluronidate may be of some importance in the functioning of the inner ear.

Continuing the work described in ref. 14, Fletcher et al. [17] measured the intrinsic viscosity (called by them "limiting viscosity") as well as specific viscosity and viscous anomalies of synovial fluids from the knee-joints of patients with rheumatoid and with osteoarthritis. The intrinsic viscosity and viscous anomaly showed no essential difference in the degree of poly-merization of the hyaluronic acid, but it is present in greater quantities in osteo- than in rheumatoid arthritis: hence the specific viscosity is higher in samples from patients suffering from the former condition, but a direct correlation is low, due to small quantities of highly polymerized hyaluronic acid.

Johnston [18] published three papers in 1955 on the viscous properties of hyaluronic acid and synovial fluids, of which the second and third are of interest here. Hyaluronic acid from synovial fluids from normal and patho-logical knee-joints was studied in both Ostwald and Couette-type visco-meters. Changes in refractive index were also related to changes in pH. Curves plotting sedimentation constants against relative viscosities differed as between rheumatoid arthritic patients and controls. The addition of the enzyme hyaluronidase to control samples produced effects similar to those found in arthritic samples. The intrinsic viscosity was found to be independent of shear rate. The best diagnostic criterion would seem to be the protein content divided by the viscosity.

The work is continued in the third paper which claims more viscous anomaly in normal than in pathological fluids. A horizontal capillary vis-cometer is described for measuring the viscous anomalies.

Barnett [19] used synovial fluid from six post-mortem apparently normal subjects (40 years old), and also from the joints of anaesthetized baboons. Thirdly, samples were aspirated from rheumatic patients. Using an emptying capillary microviscometer and constant pressures of extrusion, he got results confirming the earlier findings of Ogston and Stainer [10–12]. There is a double-logarithmic relation between the "viscosity" and the ratio of column length to time. The author suggests that an increase in this ratio and in the viscosity itself reflects the results of successful treatment. (In an earlier paper, Barnett [20] describes briefly the use of a rolling sphere viscometer for synovial fluids.) Bloch and Dintenfass [21], using a cone-cone viscometer, find considerable agreement between their results and those of Barnett.

The year 1966 marks a second turning-point in the development of research on synovial fluids. After that date, most workers were able to use rheogonio-meters or similar sophisticated instrumentation. In that year, MacConnaill [22]* wrote a most useful review article on all the earlier work, with special emphasis on the lubricating properties of synovial fluids, a subject to be discussed later.

Of particular value is his summary of a thesis by Negami [23], since this is published in a not easily accessible place. This author developed a torsion pendulum type of oscillating cylinder viscometer. He distinguishes between what he calls "static" and "dynamic" elasticity, the latter term referring to the flowing material and the former to values obtained by extrapolation to zero shear rate. The dynamic elasticity increases as viscosity falls with rising shear rate. The present author has been able to study this thesis by Negami only as described by MacConnaill, but the latter implies that these two "elasticities" correspond to the real and imaginary parts of the complex modulus. It is not quite clear why this should be so, since both "parts" are determined, without extrapolation from the ratio of maximum stress and strain and the phase angle, in a purely dynamic experiment (see Chapter II).

Negami studied variations in these moduli** with changes in temperature, pH and hyaluronic acid concentration.

Using a cone-plate rheogoniometer, Davies and Palfrey [24] compared the properties of synovial fluids from different joints of heifers. Usually, as was to be expected, the samples showed Newtonian behaviour at high shear rates but in some samples, also at low shear rates. At high shear rates only, "normal" forces (i.e. at right angles to the plane of shear) were found. The

---

* Doubtless following Dintenfass, MacConnaill uses the term "thixotropy" for "shear thinning". For the arguments against this practice, see Chapter I. Note also that Ogston's name is mis-spelt in several places in MacConnaill's paper.
** The "imaginary" part of the modulus is, of course, a measure of the viscosity for any given frequency.

importance of these properties in relation to the movement of joints is discussed.

King [25], working in co-operation with Davies and Palfrey, used a rotation type of shear and found, with samples from bullocks, good double-logarithmic curves to relate stress to shear rate over several decades. One sample showed no "normal" forces, but only three samples were studied.

Ferguson et al. [26] found a correlation between patients' subjectively assessed "joint stiffness" in the knee and the measured flow properties of the mucus. More significant is the finding that there are differences in rheological characteristics in cases of osteoarthritis, rheumatoid arthritis and traumatic effusions, thus confirming earlier findings of Myers et al. [27].

Ferguson et al. also found marked normal stresses. In a somewhat controversial paper, Theyse and Vos [28] postulated the importance of normal stresses in relation to the functioning of joints. An instrument similar to the cone-plate rheogoniometer was used.

Before we consider the lubrication aspects of the function of synovial fluids, the present section might well be closed with a brief summary of papers* which were read at a Conference of the British Society of Rheology on "Rheology in Medicine and Pharmacy", held in London in April 1970. Other papers from this Conference will be referred to later in this book.

In the first paper (P. C. Sellers, D. Dowson and V. Wright, p. 2), the authors confirm that synovial fluid is a visco-elastic liquid showing shear thinning, in both normal and pathological conditions. Some, but not all, of the viscosity–shear rate curves show double-logarithmic linearity. The hyaluronic acid concentrations, and hence the viscosity, are considerably reduced in cases of osteoarthritis. References are given to papers (most of which we have already discussed) which deal with the significance of elasticity and normal forces. "Low friction in joints is associated with a full film of lubricant which separates the surfaces and high friction results from thin films of localized boundary friction due to asperity contacts." (See also later in this chapter.) G. Nuki and J. Ferguson (p. 8) discussed the nature and significance of macromolecular complexes on fluid from normal and diseased joints, using a rheogoniometer. From experiments with hydrogen bond breaking agents, they concluded that a macromolecular complex exists under normal, but not under inflammatory conditions. Extracts of proteolytic enzymes from pathological fluids appear not to affect the flow characteristics of normal fluid. The advantages of rheological, as opposed to chemical test methods

---

* Since these papers were published in Rheol. Acta, 10: 1 (1971), they will not all be listed separately among the references for this chapter.

were stressed. Attempts to identify the macromolecular complex with the ultracentrifuge were not successful.

The same authors (but in reversed order, p. 15) used synthetic macromolecular solutions whose behaviour they compared with that of synovial fluid.

When synovial fluids are not lubricating the joints adequately, it is natural that attempts should be made to replace them with artificial lubricants. M. A. M. A. Younes, P. S. Walker, P. C. Seller, D. Dowson and V. Wright (p. 21) discussed this problem. They used polyethylene oxide and sodium carloxymethyl cellulose. The former, although a comparatively simple molecule, did not withstand mechanical working at all well. The latter, however, showed a high degree of resistance and seemed to be quite a possible substitute.

Lubrication theory and joint friction

"Tribology", the science of lubrication and friction, is on the borderlines of rheology. Perhaps the rather arbitrary distinctions, such as this and also that between rheology and hydrodynamics, will tend to disappear. Lubrication theory is, however, a highly specialized field and its application to synovial fluids will be outlined only briefly in the present book.

There is, of course, much overlapping, as we have already seen. An important paper covering both fields was published by Caygill and West [29]. These authors used a rheogoniometer, studying the various components of the stress tensor (see Chapter I) in simple shear. Four normal post-mortem samples were compared with samples from four patients with rheumatoid arthritis. Several dilutions and two temperatures (18 and 28°C) were used. Marked time dependence and shear thinning were observed. The viscosity at high shear rates did not differ significantly in the normal and pathological samples, nor did the normal force components differ. Likewise, elasticity showed no characteristic differences. However, differences were found in the curves relating apparent viscosity to rate of shear, the pathological samples being more nearly Newtonian in character. The anomalous flow was described by an equation: $\eta = \eta_\infty + A \dot{\gamma}^{-2/3}$, where η_∞ is the viscosity at very high shear rates and $\dot{\gamma}$ is the shear rate (present author's symbols). This finding may have clinical significance. The combined rolling and sliding action of the joint means that much of the joint surface is not in direct contact with the synovial fluid. Such lubrication as it provides is of the type known as "boundary lubrication". This point is discussed in a short communication by Dintenfass [30].

Some years previously, Lewis and McCutchen [31] had proposed a type of lubrication called "weeping lubrication" which MacConnaill [22] described as postulating that "the synovial fluid is contained within the articular cartilage when it is moving and pressed out from it when it is at rest". But he adds: "these views produced a lively controversy".

Recent outstanding work, already briefly mentioned, has been done by Wright and his colleagues [32–37] in Leeds. In the first of these papers, the authors quote considerable earlier work on joint stiffness, not especially concerned with synovial fluid. They propose a rheological model for stress–strain–time data obtained from the human finger joint. This model is shown in Fig. VIII-2.

Hooke   Newton   St. Venant   Kelvin

Fig. VIII (2). Model for stress–strain–time data for human finger point (Wright et al.).

"The torque required to impart a passive sinusoidal motion to the joint was recorded from strain gauges bonded to a lever and, suitably amplified, displayed on the vertical axis of a cathode ray oscilloscope." Characteristic hysteresis loops were shown relating stress to strain. The torque needed to overcome viscous resistance contributes little to overall stiffness. "The apparent correlation existing between torque and velocity has been shown to be amplitude dependent, and hence attributable to plasticity." The authors conclude with a discussion on the value and limitations of the use of "rheological models" to describe the behaviour of biological systems.

The second paper [33] extends this work to the knee-joint, which is more complex than that of the finger, and whose anatomy is discussed in this paper. An oscillatory sinusoidal motion does not involve interference by reflex or voluntary muscle action. Quite an elaborate apparatus is described and typical hysteresis loop curves are shown which relate torque to angular displacement. The stiffness increases significantly as the amplitude of displacement increases. The relationships between torque, frequency and velocity are complex, but it is concluded that the major contribution to overall stiffness is elastic. Plastic stiffness was significant. At small amplitudes "Coulomb" frictional stiffness was comparable with plastic stiffness. Viscous stiffness was small.

In the third paper to be discussed here [34] the authors describe a reciprocating friction machine with an approximately sinusoidal motion. "Model" experiments were done using a steel ball and a rubber surface. The addition of hyaluronic acid as a lubricant (to water) produced a marked reduction in friction. Calculations involving the viscosity of synovial fluid, the natural lubricant in the joints, are difficult because of the non-Newtonian character of the material.

Using neoprene rubber on a glass plate, the frictional force was found to be independent of the hyaluronic acid concentration over quite a wide range.

The lubrication of cartilage in the joint raises further complications, partly on account of its porous nature. Flat-ended specimens of cartilage were loaded on a glass plate until loss of liquid ("ring out") was complete and coefficients of friction were found to be only slightly lower than those of rubber, i.e. much higher than those found when the cartilage was soaked with synovial fluid. But in hydrodynamic lubrication, synovial fluid was little better than water; however, the nature of the lubrication process depends largely on the load on the joint. Under light loads, e.g. during the swing phase in walking, separation of the cartilage surfaces is probably great enough for hydrodynamic action.

Surface roughness of the cartilage also plays an important part, resulting in what is known as "squeeze film" friction. Pools of synovial fluid fill the "valleys" which subsequently become enriched and "gel-like" under the action of the loads.

In the two papers [35] in the "Annals of Rheumatic Diseases", the authors summarize the earlier work and make a more detailed study of the microscopic surface on the cartilage. Measurements were made using a diamond point and also an electron microscope. A hitherto unexplored type of lubrication is also discussed ("fluid enrichment").

In a letter to "Nature" [36] more is said about "boosted lubrication". When synovial fluid is squeezed between surfaces it flows in preferred directions. (This is, perhaps, not surprising, in view of its anisotropic characteristics). A comparison is made with the behaviour of synthetic polymers.

The last paper by Wright to be discussed here [37] was a contribution to the British Society of Rheology's Conference on "Rheology in Medicine and Pharmacy" already referred to. Much of the earlier work is summarized, but also reflex muscle activity is measured at the joint, e.g. in myotonia congenita and in Parkinson's disease. In patients without neuromuscular disease, the muscles are electrically inactive during sinusoidal motion whereas active contraction distorts the tracing. This finding may be of importance in distinguishing psychosomatic factors. Not unnaturally, stiffness increases with advancing age and also with cold conditions. "Articular gelling in patients with osteo-arthritis may be demonstrated after periods of rest and the results suggest that such changes are due to alterations in the periarticular and muscle tissues."

It has not been possible in a book of this kind to do full justice to the work of the Bioengineering Group at Leeds. In the present author's opinion, however, this body of work represents the most thorough study of joint friction and of the part played by synovial fluids in its reduction in health and disease as yet undertaken.

We have left until last, in this chapter, the work of Hwang et al. [38] because these authors deal with several types of mucus, which will be discussed elsewhere in this book.

The rheometer especially designed for this work consisted of a very small nickel sphere (diameter 18–50 μ) which is moved horizontally by means of a magnetic field, through the mucus to be tested. (This is essentially the same as the method described by Behar and Frey [39] whose work was evidently not known to Hwang. He reminds us, however, that Freundlich and Seifriz [40] proposed a not dissimilar method as early as 1923.) The sphere can be oscillated sinusoidally but the present paper is concerned mainly with direct pull. A rheogoniometer was also used.

For experimental purposes, a vegetable gum was used in place of a mucus. The dynamic viscosities could be calculated from data from both the microsphere instrument and the rheogoniometer and good agreement was found between them, except when very short times were required for the displacement of the sphere. Cervical and bronchial mucus were those principally tested. The authors pointed out the importance of Reiner's "Deborah number" in relation to experimental conditions and this was found to be of the order of 2000. (This is the ratio of relaxation time to experimentation time.) The distance between the molecular entanglements differed for different types of mucus.

REFERENCES

[1] Schneider, J., *Biochem. Z.*, 160: 325 (1925).
[2] Panizza, B., *Arch. fisiol.*, 29: 576 (1930–1931).
[3] Bywaters, E. G. L., *J. Path. Bact.*, 44: 247 (1937).
[4] Chain, E. and Duthie E. S., *Nature*, 144: 977 (1939) and *Brit. J. exp. Path.*, 21: 324 (1940).
[5] Davies, D. V., *J. Anat.*, 78: 68 (1944); also Davies, D. V. and Palfrey, A. J., *Soc. chem. Ind. Monogr.*, No. 24 (1966).
[6] Blix, G. and Snellman. O., *Arkiv Kemi Mineral. Geol.*, 19A (No. 32): 1 (1945) (in English).
[7] Ragan, C., *Proc. Soc. exp. Biol. Med.*, 63: 572 (1946).

[8]   Ropes, M. W., Robertson, W. B., Rossmeisl, E. C., Peabody, R. B. and Bauer, W., *Acta med. scand.*, 128, suppl. 196: 700 (1947).

[9]   Rimmington, C., *Ann. rheum. Dis.*, 8: 34 (1949).

[10] Ogston, A. G. and Stanier, J. E., *Biochem. J.*, 46: 364 (1950) and 49: 585 (1951).

[11] Ogston, A. G. and Stainer, J. E., *J. Physiol.* (Lond.), 119: 244 (1953) and 119: 253 (1953).

[12] Ogston, A. G., Stainer, J. E., Toms., B. A. and Strawbridge, D. J., *Nature*, 165: 571 (1950).

[13] Sundblad, L., *Acta. Soc. Med. Upsalien.*, 58: 113 (1953) (in English).

[14] Scott Blair, G. W., Williams, P. O., Fletcher, E. T. D. and Markham, R. L., *Biochem. J.*, 56: 504 (1954).

[15] Jensen, C. E. and Koefoed, J., *J. Colloid Sci.*, 9: 460 (1954).

[16] Jensen, C. E., Koefoed, J. and Vilstrup, T., *Nature*, 174: 1101 (1954).

[17] Fletcher, E., Jacobs, J. H. and Markham, R. L., *Clin. Sci.*, 14: 653 (1955).

[18] Johnston, J. P., *Biochem. J.*, 59: 626 and 633 (1955).

[19] Barnett, C. H., *Ann. rheum. Dis.*, 17: 229 (1958).

[20] Barnett, C. H., *Phys. Med. Biol.*, 1: 380 (1957).

[21] Bloch, B. and Dintenfass, L., *Aust. N. Z. J. Surg.*, 33: 108 (1963).

[22] MacConnaill, M. A., *Lab. Pract.*, 15: 60 (1966).

[23] Negami, S., *Dynamic Mechanical Properties of Synovial Fluid*, Lehigh Univ. Thesis, Bethlehem, Pa., 1964.

[24] Davies, D. V. and Palfrey, A. J., *Soc. Chem. Ind. Monogr.*, No. 24 (1966).

[25] King, R. G., *Rheol. Acta*, 5: 41 (1966).

[26] Ferguson, J., Boyle, J. A., McSween, R. N. M. and Jasani, M. J., *Biorheology*, 5: 119 (1968).

[27] Myers, R. R., Negami, S. and White, R. K., *Biorheology*, 3: 197 (1966).

[28] Theyse, F. H. and Vos. R., *Lubric. Engng*, 26: 101 (1970).

[29] Caygill, J. C. and West, G. H., *Med. biol. Engng*, 7: 507 (1969).

[30] Dintenfass, L., *Nature*, 197: 496 (1963).

[31] Lewis, P. R. and McCutchen, C. W., *Nature*, 184: 1284 (1959).

[32] Johns, R. J. and Wright, V., *Biorheology*, 2: 87 (1964).

[33] Goddard, R., Dowson, D., Longfield, M. D. and Wright, V., *Rheol. Acta*, 8: 229 (1969).

[34] Walker, P. S., Dowson, D., Longfield, M. D. and Wright, V., *Rheol. Acta*, 8: 234 (1969).

[35] Walker, P. S., Dowson, D., Longfield, M. D. and Wright, V., *Ann. rheum. Dis.*, 27: 512 (1968).

[36] Walker, P. S., Sikorski, J., Dowson, D. and Wright, V., *Nature*, 225: 956 (1970).

[37] Wright, V., Dowson, D. and Goddard, R., *Rheol. Acta*, 10: 61 (1971).

[38] Hwang, S. H., Litt, M. and Forsman, W. C., *Rheol. Acta*, 8: 438 (1969).

[39] Behar, Y. and Frey, E. H., *Bull. Res. Counc. Israel A*, 5: 82 (1955).

[40] Freundlich, H. and Seifriz, W., *Z. phys. Chem.*, 104: 233 (1923).

BOOK FOR FURTHER READING

Barnett, C. H., Davies, D. V. and MacConnaill, M.A., *Synovial Joints* (Longman, London, 1961).

Chapter IX

RHEOLOGY OF SPUTUM, BRONCHIAL MUCUS AND SALIVA

In vitro experiments

Although this book is not written primarily for rheologists, but rather for biologists, medical practitioners and others who wish to have some ideas as to how rheology can help in their work, nevertheless a very brief account of the functions of bronchial mucus, although superfluous for biologists, should perhaps serve as an introduction to this chapter. This is largely taken from an account given by two rheologists, Merrill and Wells [1], in an article which also deals with many other biological "fluids".

The lung is very much like an inverted tree (the "bronchial tree"). Air from the pharynx passes down the main trunk and hence into the bronchi and bronchioles ("branches and twigs") and finally to the alveoli and atria which correspond roughly to the leaf tissue in the tree.

The bronchial mucus is produced by glands and the lower part of the system and it works upwards, propelled by the action of hair-like "cilia". The function of the mucus is to keep the passages both clean and moist. The evaporation of the moisture helps to control the temperature of the body. In bronchitis and similar diseases, the rheological properties of the mucus are affected in such a way as to inhibit these activities.

Bronchial mucus is not easy to collect in sputum, unmixed with saliva. Samples are generally not homogeneous (as is also the case with cervical mucus) and, in rheological testing, a balance must be struck between the risks of getting a non-representative sample and of "mixing" the sample in such a way as to destroy the very structure that is required to be measured.

Bronchial mucus has long been known to show shear thinning, and Merrill and Wells suggest that during the forward motion of the cilia, at lower shear rates, high consistency conditions preponderate, whereas on the backward stroke, the mucus "thins", thus accounting for its motion upwards against gravity, in an upright person.

If the consistency of the mucus becomes too high, the patient has difficulty in breathing and, as we shall see later, various "softening" agents have been proposed to give relief.

As in earlier chapters, a somewhat arbitrary selection of the voluminous literature will be made, hoping to give the reader an overall idea of the history and present state of knowledge concerning sputum rheology. Even more so than has been the case with cervical and synovial secretions, there is much replication in the literature, doubtless because of the very wide range of journals in which the work has been published.

The present author has not found much of importance before 1950 and we will follow, very approximately in chronological order, some subsequent researches.

In that year, Armstrong and White [2] published a short note on sputum viscosity, and in 1953 Elmes and White [3] wrote an important paper extending this work, in which they showed that the enzyme deoxyribonuclease reduces the consistency (or "viscosity", as they called it) of purulent bronchitic sputum. Two types of rheometer were used: a falling sphere (modified from a design described by Lucas and Henderson [4] much earlier) and also a falling perforated disk. Most of the experiments were done with the latter. Elmes and White gave a summarized account of their work at the 2nd International Congress on Rheology [5].

Chronic bronchitis is characterized by four principal conditions: (1) thickening of the lining of the bronchi; (2) destruction of ciliated cells; (3) increase in the amount of secretion; and (4) loss of rigidity of the walls of the bronchi. At the time of writing, although the use of antibiotics could control the infection, the amount of mucus might continue to be so excessive as to obstruct the bronchi and even to result in death. The authors showed electron- and phase-contrast micrographs of purulent mucus showing coarse bundles of parallel fibres, which are birefringent. After the infection is over, "phase-contrast microscopy of wet smears shows few pus cells, but the presence of a fibrillary structure is indicated by the arrangement of particulate matter in lines".

The authors stress the difficulties in getting adequate reproducible and homogeneous samples for rheological testing. Following unsatisfactory results with the sphere viscometer, the perforated disk method was found to be much more satisfactory. The time was measured for a fixed volume to be forced through the holes under a standard load. The mucus is softened by repeated shearing, especially in the case of pathological samples. A modification of the disk rheometer, suitable for these pathological samples, was made, using a standard hypodermic syringe held vertically and fitted with a perforated brass block cylinder. The nozzle was blocked, and the mucus was forced through the perforations by loading the piston. A kind of yield vaule was measured and this was shown to increase during the earlier stages of a bronchial attack. During the purulent phase, the consistency could be much reduced by the administration of deoxyribonuclease on alternate days, but it "had no obvious effect on the clinical course of the disease". We shall see how later work extended and developed the use of enzyme "softeners" of mucus.

The next papers which seem worthy of note are those of Blanchard [6]. This author gives a summary of the general principles of rheology before describing his own experiments on bronchial mucus, using a rotating cone-plate viscometer. He confirms the findings of Elmes and White and finally concludes that "the viscosity of sputum is due to the internal friction of fibres of mucoprotein and deoxyribonucleoprotein; it behaves in a non-Newtonian fashion, showing a yield value and a fall of consistency at higher rates of shear. Sputum produced overnight is more viscous than that produced during the day; the viscosity is influenced by the state of bodily hydration." In the second paper, Blanchard [7] discussed the effects of operations under general anaesthesia on the properties of the mucus.

In the later 1950's, there was much discussion in the medical journals on the relative merits of various enzymes and of detergents in the form of aerosols, in the treatment of bronchitis. (An Editorial in the "British Medical Journal" of March 31, 1962 gives a good summary of the position at that time.)

Palmer [8] pointed out that trypsin, sometimes used, is an irritant and a foreign protein, involving possible risks of malignancy. "Alevaire" is a detergent aerosol and this was tried for comparison with a sodium bicarbonate solution, without detergent, on 25 patients. Both treatments help with expectoration; but it is concluded that, since the effects appear to be due simply to the hydration of the sputum, a simple water aerosol should be equally effective.

Robinson et al. [9] treated 36 inpatients with chronic bronchitis by insufflation into the peripheral bronchial tree of powdered deoxyribonuclease and chymotrypsin. These enzymes reduced the consistency of the mucus and 67% of the patients considered the treatment helpful. It was suggested that a more homogeneous powder, combined with a bronchodilator, might give even better results. This and other treatments were followed up in the second part of this paper.

In the same year, Hillis [10] did a statistically well-controlled experiment, using potassium iodide, ammonium chloride, ipecacuanha, and a placebo and found no significant effects on the consistency of the sputum, except in the case of KI. But the mucus was first homogenized and, although the authors suggest that this homogenization may have been inadequate, it seems more likely, on the contrary, that the homogenization destroyed much of the structure that the authors were trying to measure.

Soon after this, White et al. [11] and White and Elmes [12] published further work on this subject. In the first short note, they described measurements of the intrinsic viscosity of bronchial mucus and in the second article, the effects of "homogenization" and subsequent resting are discussed. The authors still felt that the yield value was the best rheological property to use as a criterion for the fibrous structure. Homogenization in water and in various strengths of NaCl show a breakdown mainly dependent on disintegration of the mucoprotein. The consistency is partially recovered on resting, but the primary structure is never fully regained. There are difficulties, however, in comparing the properties of the original mucus with those of the isolated mucoprotein because the latter is inevitably changed irreversibly in the process of preparation. The intrinsic viscosity of the mucoprotein was determined, fairly good straight lines being obtained by plotting reduced viscosity against concentration. The aqueous preparation showed the widest scatter and the highest $[\eta]$. Solutions in 0.13 M NaCl and in 8.1 M LiBr gave much lower values and the reduced viscosities were almost independent of concentration. The fibrillar structure of the mucus is clearly seen in electron micrographs.

There was a lengthy discussion following the reading of this paper. Blanchard again stressed marked increases in "viscosity" (consistency) during surgery in bronchial cases and other changes taking place in the post-operative period. M. H. Knisely pointed out that, following deaths from asthma, the bronchial tree is sometimes found to be full of a thick, rubber-like mucus and R. D. Harkness asked to what extent the mucus had been contaminated with saliva. P. C. Elmes admitted that saliva had been present, but said that it tended to form a surface coating on the masses of sputum and probably had little effect on its consistency.

Wells et al. [13], who designed the "Wells–Brookfield cone-plate viscometer" (see Chapter II), took samples of mucus from patients with chronic obstructive emphysema and bronchitis. Very thick samples were obtained, and these were cut with scissors and placed between two pegged plates. Elasticity and shear thinning were observed.

Meanwhile, in medical circles, studies on the effectiveness of aerosol treatment in bronchial conditions continued (see Coriminas-Beret [14]) and O'Donnell and Gerachty [15] found that the consistency of mucus from bronchial patients could be much reduced in vitro by the addition of hydrogen peroxide.

In 1962, there was considerable controversy as to whether the heterogeneity of sputum and the destructive effects of mixing before rheological testing, make consistency measurements hardly worthwhile (see Lloyd-Bisley, and Bruce and Quinton [16, 17].) The former quotes the present author as saying that "the problem is, in its very nature, insoluble". In a sense, this is true; but it is perhaps not necessary to be too pessimistic. It is evident from the literature that tests on lightly mixed samples are far from worthless as are also tests, such as those of Bruce and Quinton, in which the sample is carefully separated into a thick and a thin portion and tests are made only on the former. Bruce and Quinton measured the torque on a disk resting on the sample, which was placed on a rotating parallel plate and found that oral administration of α-chymotrypsin reduced the consistency of the thick part of the sputum.

Bürgi [18] reported that mucolytic agents such as "Bisolvan" effectively reduce the consistency of mucus but he did not describe specific rheological tests.

The work of Hwang et al. [19] was discussed to some extent in the last chapter. The magnetic moving sphere viscometer is not very suitable for studying bronchial mucus, which is generally too opaque for the sphere to be visible. Only one sample appears to have been tested and that had been stored (frozen) for 2 years! All that can be said is that such tests as were possible confirm the general picture of bronchial mucus as a visco-elastic system.

Davis and Dippy [20] have studied sputum from chronic bronchitis separated as far as possible from saliva. They used a cone-plate rheometer which was continuously accelerated and decelerated. Fig. IX-1 shows a typical flow curve. The structure is broken down in shear and recovers slowly on standing. Initially, there is a yield value.

Using a modified rheogoniometer, the authors also obtained a typical creep curve, followed by a delayed elastic recovery. A Burgers model (mistakenly called a Voigt model) describes this behaviour*.

Measuring the visco-elastic properties with the oscillating rheogoniometer (cone-plate) the authors conclude that the structure of mucus is comparable to that of uncross-linked polymer solutions.

In the Conference of the British Society of Rheology on "Rheology in Medicine and Pharmacy" (1970) already referred to in an earlier chapter, there were two papers on mucus and one on saliva. It will be convenient to consider them together.

Sturgess et al. [21] studied sputum using a rheogoniometer with oscillation

---

* A dashpot and spring in parallel attached to a dashpot and spring in series.

for lower shear rates and a cone-plate rotation viscometer for higher shear rates. The samples, collected over a short time, were tested within 2 or 3 hours of collection and without pre-treatment. At high shear rates (up to 2000 s^{-1}), the first shearing progressively breaks down the structure and subsequent shearings give almost identical straight lines, curving only at the bottom so that there is an extrapolated intercept on the stress axis (approximately a Bingham system). Duplicates from the same original sample give almost identical curves. With the rheogoniometer, if viscosity is plotted at ascending shear rates, there is a flat or irregular portion of the curve at about the middle of the range; but this is absent when ascending shear rates are plotted. The "plateau" is found in both mucoid normal and purulent samples. A possible qualitative explanation, in terms of the strength of molecular linkages, is given. The "plateau" is also present in the elasticity curves.

Comparing healthy with purulent sputum, the authors find that, although the latter is more easily coughed up, it has, in fact, a higher initial viscosity; but the structure is more easily broken down. This is especially true in cases of cystic fibrosis, (often called "mucoviscidosis"), in which the sputum is, in fact, initially no more viscous than is that of bronchitis. Effects of storing sputum samples and of varying the test temperatures are also discussed. References to earlier work by Sturgess are given.

The other paper on sputum, by Tekeres and Geddes [22], also describes experiments with a similar cone-plate viscometer, and the data, though less numerous, are not dissimilar. The samples studied were from chronic bronchitics and it was found that the breakdown produced by shearing was recovered on subsequent resting.

Davis [23], in his contribution to the Conference, compares the properties of sputum discussed in his earlier paper [20] with those of saliva. Conditions in which saliva is excessively plentiful and, in other cases, having an abnormally high viscosity are known. Whether there is a correlation between viscosity and mucin content appears to be in dispute: Bagnall and Young [24] are quoted as finding no correlation.

Davis claims that the ordinary rotating cone-plate viscometer is inadequate for the study of saliva, which is a visco-elastic system, and after preliminary tests with the cone-plate, he uses only a rheogoniometer in oscillation. The variations of both the real and imaginary (viscous) parts of the complex modulus with frequency give good straight lines plotting double logarithms. In general, the rheology is very similar to that of sputum: like sputum, the properties of saliva vary from day to day. Storage produces a fall in consistency, mainly caused by the action of enzymes on sialic acid. Mucolytic agents act on saliva in much the same way as they do on sputum.

In the Discussion, several speakers stressed the fact that, in the mouth, rates of shear are generally high (see Chapter XI of this book); but the speakers particularly stressed the more anomalous and rheologically interesting behaviour at very low shear rates.

At the 5th International Congress on Rheology, Nagaoka and Fukushima [25] described work in which they had separated sputum from saliva by expelling the sputum onto a square of gauze, which absorbs the saliva. The sputum was sucked into a glass tube connected to a syringe to produce the suction. The negative pressure was recorded on an electric manometer. The volume change in the air space produced by the advancing column of sputum was also recorded. The authors found, as did many other workers, that the sputum showed shear thinning following a yield value. They follow Merrill and Wells [1] rather closely in suggesting that, in chronic bronchitis, "the force delivered by the cilia by the rapid forward stroke to the surrounding mucus falls short of the elastic limit, and during the longer interval of the recovery stroke, elasticity may cause a return of the stretched mucus to the original position". The authors also used a perforated disk apparatus similar to that of White and Elmes [12] to measure yield values. An apparatus was also made to produce sudden pressures of about the magnitude of those which occur in coughing.

The thinner part of the sputum is described as thixotropic, more or less complete recovery taking place after about half an hour.

Clinically, perhaps the most valuable paper yet published is that of Charman and Reid [26]. These authors took sputum from 77 patients with chronic bronchitis (24), bronchiectasis (15), asthma (19) and cystic fibrosis (19). (The numbers in parentheses refer to the number of patients.) Ages ranged from 6 to 75: 26 samples were mucoid (i.e. devoid of pus), 22 mucopurulent (partly purulent) and 29 purulent (permeated by pus). The diagnoses of the various conditions were made by standard methods that need not be described here. The usual type of cone-plate viscometer was used, in which there is a uniform acceleration over a period of time, followed by a uniform deceleration. Typical hysteresis loops were obtained, quite similar to those already described from other workers (see Fig. IX-1). Three ranges of shear rate were used: $0-180$ s^{-1}, $0-1800$ s^{-1} and $0-18000$ s^{-1}. Tests were made at room temperature. The lowest part of the curves was found, by standardization with a Newtonian fluid, to be subject to inertia errors.

The logarithms of the viscosities were plotted against the logarithms of the shear rates, to give a series of curves as shown in Fig. IX-2. The slope was measured over the lower (linear) part of each curve as shown by the tangents. The rates of shear used to calculate apparent viscosity were taken three-quarters way through the curves, i.e. at 135, 1350 and 13,500 s^{-1}.

124

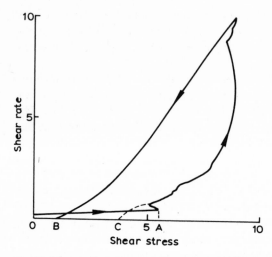

Fig. IX (1). A typical rheogram for sputum. (A) Static yield value. (B) Dynamic yield value. (C) Extrapolated static yield value. Shear rate: 10 units = 1692 s^{-1}; shear stress: 10 units = 980 dynes/cm^2. Reproduced from Davis, S.S. and Dippy, J.E., Biorheology, 6: 13 (1969).

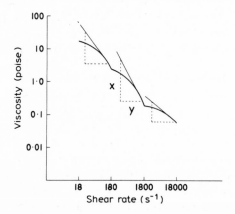

Fig. IX (2). Bronchial mucus: Variation in viscosity with shear rate, both plotted on a logarithmic scale, for speed ranges 0–180, 0–1800 and 0–18000 s^{-1}. The rate of fall of viscosity i.e., the "slope", is calculated by dividing X by Y. Reproduced from Charman, J. and Reid, L., Biorheology, 9: 185 (1972).

Apparent viscosities were calculated at the point of shear thinning and at $900 s^{-1}$. The area under the hysteresis loop was taken as a measure of "breakdown". It was found adequate to use the same aliquot for all three shear ranges.

It is found that it is possible to "store" mucus, for comparatively short periods under certain conditions of freezing and thawing, but it seemed best to use fresh samples. Viscosity itself is not very helpful for diagnosis though asthmatic patients' samples seem to reach rather higher levels than do those of the other conditions studied. Cystic fibrosis samples were significantly less viscous than those from chronic bronchitis and bronchiectasis.

It is concluded that the viscosity depends more on whether the sample is mucoid or purulent than on the nature of the disease, the former being generally more viscous than the latter.

The work of a number of other authors (some not mentioned in this chapter) is quoted, but the conclusions were very diverse, mainly because too few patients were tested to allow an adequate statistical analysis.

At the 6th International Congress on Rheology, held at Lyon in September 1972, Molina et al. [27] read a paper on the rheology of a very large number of samples of sputum. More than 15,000 determinations of the consistency of bronchial mucus were made using a cone-plate microviscometer. The "aqueous phase" (saliva?) was first removed. The authors claim to have found "a correlation between rheological data and histological and biochemical modifications of bronchial mucus"*. It is further concluded that rheological studies are of value in assessing the effectiveness of mucolytic treatments, and that the combined use of mucolytics and antibiotics is more effective than that of either treatment alone.

Experiments on animals

The use of animals, and especially the rat (though dogs and guinea pigs and cats have also been used) makes it possible to study the flow of mucus while still in the trachea, and bronchial conditions can be induced by the action of such irritants as sulphur dioxide. Information on this subject is taken from a paper by Lightowler and Williams [28]. Earlier work is quoted by Dalhamn [29] and by Dalhamn and Reid [30] but Lightowler and Williams regard their results as inconclusive. However, results reported in a later paper by Prof. Lynne Reid [31], obtained by Quevauviller and Huyen [32], and by Hanasato [33] on the histological changes produced by SO_2, are

* Author's translation from the French summary.

confirmed by the authors. Most of this paper is concerned with histological changes: only the results that bear on the rheological properties of the mucus will be briefly discussed here.

Several authors have measured the rate of flow of mucus in the trachea of various species. Even for the same species, there are wide variations. Lightowler and Williams' results agree most closely with those of Dalhamn [29], giving a mean speed of 12 mm/min. "Dalhamn reported marked functional changes with cessation of mucus flow in rats exposed to 10 p.p.m. SO_2 for periods up to 67 days. He suggested that the thickness of the mucus blanket was an important factor in slowing of mucus flow, whilst Florey et al. [34]* observed increased flow rates with increased mucus production."

In treated rats, the mucus appeared to divide into two layers after prolonged treatment. There was a lower watery layer, flowing rapidly, and an upper, thick, opaque mucus that remained almost stationary. The cilia vibrate in the lower viscosity fluid. (Hilding [35] had already pointed out the importance of having a more or less homogeneous mucus which can be moved by the vibrating cilia.)

The animals had not recovered even 2 months after the end of the treatment with SO_2.

Dalhamn and Reid [30] found no appreciable change in the beat rates of the cilia in rabbits exposed to a mixture of ammonia and carbon in air.

Lightowler and Williams' final conclusion is that there is a reduction in the speed of flow of mucus in the exposed trachea which increases with the severity of the condition. This is due to changes in the quantity and quality of the mucus which affect the extent to which the cilia are able to move the mucus.

(Other relevant papers are listed in refs. 36–38.)

---

* The reference numbers are adjusted to the present text.

REFERENCES

[1]   Merrill, E. W. and Wells, R. E., *Appl. Mech. Rev.*, 14: 663 (1961).
[2]   Armstrong, J. B. and White, J. C., *Lancet*, 259: 739 (1950).
[3]   Elmes, P. C. and White, J. C., *Thorax*, 8: 295 (1953).
[4]   Lucas, G. H. W. and Henderson, V. E., *Arch. int. Pharmacol.*, 54: 201 (1963).
[5]   Elms, P. C. and White, J. C., *Proc. 2nd int. Congr. Rheol.*, Oxford, 1953 (Ed. V. G. W. Harrison), p. 382 (Butterworth, London, 1954).
[6]   Blanchard, G., *Arch. Middx. Hosp.*, 5: 222 (1955).
[7]   Blanchard, G., *Dis. Chest.*, 37: 75 (1960).

[8] Palmer, K. N. V., *Lancet*, March 23, 1957, p. 611 [see also Palmer, K. N. V., Bal-
     lantyne, D., Diament, M. L. and Hamilton, W. F. D., *Brit. J. Dis. Chest*, 64: 185
     (1970)].

[9] Robinson, W., Woolley, P. B. and Altounyan, R. E. C., *Lancet*, Oct. 18, 1958,
     pp. 819 and 821.

[10] Hillis, B. R., *Scot. med. J.*, 3: 252 (1958).

[11] White, J. C., Elmes, P. C. and Whitley, W., *Nature*, 183: 1810 (1959).

[12] White, J. C. and Elmes, P. C., in *Flow Properties of Blood and Other Biological
     Systems* (Ed. A. L. Copley and G. Stainsby), p. 259 (Pergamon Press, Oxford, 1960).

[13] Wells, R. E., Denton, R. and Merrill, E. W., *J. Lab. Clin. Med.*, 57: 646 (1961).

[14] Coriminas-Beret, C., *Brit. med. J.*, No. 5294, June 23, 1962, p. 1763.

[15] O'Donnell, J. and Gerachty, M., *Brit. med. J.*, No. 5293, June 15, 1962, p. 1695.

[16] Bruce, R. A. and Quinton, K. C., *Brit. med. J.*, No. 5274: 282 (Feb. 3, 1962) and
     reply to Lloyd-Bisley, B., *Brit. med. J.*, No. 5281: 801 (Mar. 24, 1962).

[17] Lloyd-Bisley, B., *Brit. med. J.*, No. 5281: 801 (Mar. 24, 1962).

[18] Bürgi, H., *Lancet*, No. 7360: 644 (Sept. 19, 1964).

[19] Hwang, S. H., Litt, M. and Forsman, W. C., *Rheol. Acta*, 8: 438 (1969).

[20] Davis, S. S. and Dippy, J. E., *Biorheology*, 6: 11 (1969).

[21] Sturgess, J., Palfrey, A. J. and Reid, L., *Rheol. Acta*, 10: 36 (1971).

[22] Tekeres, M. and Geddes, I. C., *Rheol. Acta*, 10: 44 (1971).

[23] Davis, S. S., *Rheol. Acta*, 10: 28 (1971).

[24] Bagnall, J. S. and Young, E. G., *J. dent. Res.*, 10: 393 (1930).

[25] Nagaoka, S. and Fukushima, Y., in *Proc. 5th int. Congr. Rheol.*, 1968 (Ed. S. Onogi),
     Vol. 2, p. 153 (Tokyo Univ. Press, 1969).

[26] Charman, J. and Reid, L., *Biorheology*, 9: 185 (1972).

[27] Molina, C., Aiache, J. M., Brun, J. and Chemist, J. C., *J. Biorheol.*, (abstract) 9: 160
     (1972).

[28] Lightowler, N. M. and Williams, J. R. B., *Brit. J. exp. Path.*, 50: 139 (1969).

[29] Dalhamn, T., *Acta. physiol. scand.*, 36, Suppl. 123 (1956).

[30] Dalhamn, T. and Reid, L., *Inhaled Particles and Vapours*, Vol. 2, p. 299 (Pergamon
     Press, Oxford, 1965).

[31] Reid, L., *Brit. J. exp. Path.*, 44: 437 (1963).

[32] Quevauviller, A. and Huyen, V. N., *C.R. Soc. Biol.*, 160: 1845 (1966).

[33] Hanasato, S., *Shinshu med. J.*, 15: 104 (1966).

[34] Florey, H., Carleton, H. M. and Wells, A. Q., *Brit. J. exp. Path.*, 13: 269 (1932).

[35] Hilding, A. C., *Arch. Otolaryng.*, 15: 92 (1932).

[36] Keal, E., *Poumon*, 26: 25 (1970) and 2: 51 (1970).

[37] Reid, L., *Proc. roy. Inst. Gt. Br.*, 43, No. 202: 438 (1970).

[38] Keal, E., *Postgrad. med. J.*, 47: 171 (1971).

Chapter X

THE EYE: THE LENS, INTRAOCULAR FLUID; GLAUCOMA AND CATARACT; TONOMETRY; VITREOUS HUMOUR. OTHER BODY FLUIDS. MISCELLANEOUS

The lens*

As early as 1849, Bowman [1] observed the elasticity of the lens capsule of the eye. If this was punctured, after imbibing water, it contracted rapidly, expelling the water. Eisler [2], however, showed that the elastic recovery was incomplete. Fisher [3] quotes later authors who came to somewhat contradictory conclusions concerning the elasticity of the lens; and it was this fact that led him to undertake his own extensive researches.

In the first of the papers quoted here, Fisher determined the elasticity of the human lens at birth and throughout life. The factors specifically measured were: lens thickness, stress–strain relations and Poisson's ratio. An ingenious apparatus for making these measurements is described. It is found that the thickness of the anterior capsule increases during the first 60 years of life and subsequently decreases. (The capsule is the cover that contains the lens.) The material of the capsule was found to be nearly "incompressible", i.e. does not change its volume under stress (Poisson's ratio 0.47 as compared with 0.50 for a completely incompressible material), and this did not appear to change with ageing. The relation of volume (corrected for porosity) to pressure was found to be linear, i.e. Hooke's law was obeyed. Young's modulus (of extension) fell progressively from childhood to extreme old age: from 6×10^7 to 1.5×10^7 dynes/cm². The ultimate tensile stress (at rupture) fell, with ageing, from $2.3.10^7$ to $0.7.10^7$ dynes/cm². The maximum elongation was 29%, independent of age.

Fisher compares the Young's modulus and maximum extension of the human lens capsule with values obtained by other authors for rat collagen and for the elastic tissue of the human aorta, and is led to suppose "that the human lens is surrounded in turn by tissue of the same thickness as its own

---

* The author is indebted to Dr. R. F. Fisher for copies of his very valuable papers. Most of the references to the earlier work are taken from these.

capsule, but with the properties of collagen and elastic tissue". Knowing the changes in the diameter of the lens when accommodation relaxes (from Finsham [4]), Fisher goes on to conclude "that the human lens capsule has properties intermediate between collagen and elastic tissue which produces an effective compromise between excessive lens moulding pressure associated with small elongation, and insufficient force coupled with great elongation".

The ageing processes described correspond, in many respects, with the known losses in elasticity of skin and other tissues. Vogt [5] showed that loss of accommodation with ageing is associated with loss of lens capsule elasticity; but the increase in the thickness of the capsule makes some compensation for this. In his second paper, Fisher measures the amount of stored energy in the capsule, released during accommodation. This is done by an ingenious method in which the lens is spun around its axis so that variable centrifugal forces can be applied, imitating the mechanical tensions to which it is subjected in normal vision. High speed photographic recording is required. In these experiments, the Hookean range of elasticity is exceeded and visco-elastic properties appear. The lens is isotropic in youth and old age, but anisotropic in middle life. A comparison is made with the rather different properties of the lenses of cats and rabbits.

Fisher comments that earlier workers (e.g. Fukada [6]) have shown that the crystalline lens "behaves as a rheological body". This is rather a strange expression, since the term "rheology" is always taken to include the study of elastic behaviour, in spite of its Greek derivation. But Fisher would appear to mean that "it continues to deform at a decreasing rate so long as a force is applied". This would, of course, apply to a Kelvin model of a viscous and an elastic element in parallel.

Fig. X-1 shows the profile of a young adult lens, stationary (above) and spinning at 1600 r.p.m. (below). For a very interesting discussion on the effects of lens anisotropy on accommodation, comparing man, cat and rabbit, the reader is referred to the original paper. Changes in the anisotropy with ageing are also shown as related to the loss of accommodation.

Dehydration of the nucleus of the lens leads to crystallization and, like all long-chain molecules, the protein molecules show a marked rise in Young's modulus, though Fisher stresses that the quantitative changes in moisture content of the human lens with ageing need further research.

The rise in Young's modulus in early life requires a different explanation. François et al. [7] have shown that, in early life, there is a tendency towards a fall in the molecular weight of the proteins which would tend to reduce Young's modulus.

In a fourth paper, Fisher [8] discusses lens fibre stress and the formation

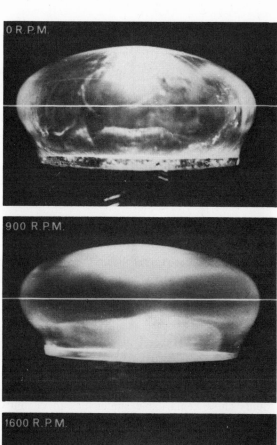

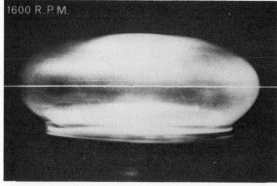

Fig. X (1). Photographs of the anterior profile of a human lens (age 21) with anterior pole uppermost. (A), stationary lens; (B), the same lens spinning at 1000 rev./min.; (C), the same lens stationary after the capsule was cut around the lens equator. (Equator shown in every case by a thick white line while a thin white line passes through the anterior pole of the lens shown in (A).) Note: Decrease in anterior polar thickness of the spinning and cut capsule profiles compared with the stationary profile. Reproduced from: Fisher, R.F., J. Physiol., 228: 765 (1973) (Plate I).

of opacity in senile cataract. Previously, nutritional, genetic and enzymic theories were widely held to be responsible; but Fisher discusses the importance of accommodation stresses in stretching the lens beyond its elastic limit. It is significant that the age of maximum accommodation (48 years) corresponds rather closely to the onset of opacities, though these are found initially mainly at the extreme edge of the lens, so that cataract is not usually observed until later in life. Fisher quotes earlier workers who suggested this cause of cataract in later life, but they did not have the advantage of the very impressive rheological data obtained by him.

Woo et al. [9] discuss stress–strain relations for the cornea and give a good bibliography of papers on elasticity of the cornea.

Cataract and glaucoma

The author, who suffers from both cataract and glaucoma, is well aware of the prevalence of these conditions. In glaucoma, the pressure within the eye is raised as a result of blocking of the ducts which allow for the outflow of the intraocular fluid. Only by making fresh ducts can total blindness be avoided if the condition is at all serious.

An early review article by Amsler and Huber [10] will give the reader a good summary of work before 1952 and it also stresses the relationship between the intraocular fluid and the cerebrospinal fluid. The eye and the brain are embryologically connected, and even after the eye is independently fully developed, the two fluids show remarkable similarities in chemical composition and rheological behaviour. This article was written 20 years ago and in any case, it was more concerned with the physiology than with the rheology of the fluids, though the importance of changes in the viscosity of intraocular fluid under certain pathological conditions was stressed. Under physiological conditions, the fluid is probably so nearly Newtonian that rheological studies may not throw much light on its behaviour. At any rate, the present author has not been able to find much published work on the rheology of either cerebrospinal or intraocular fluid.

Tonometry

In the diagnosis of glaucoma, and in the control of the disease, an accurate measure of the internal pressure is needed. Rather crude "rheometers" were commonly used for this purpose. A small weight was lowered onto the anaesthetized eyeball and the displacement was read off on a scale, e.g. see Friedenwald and Moses [11].

In recent times, however, much more sophisticated techniques have become available. When the contact points of the modern "tonometer" touch the eyeball, the patient is hardly conscious of the contact and the time of contact can be so short that the initial pressure can be recorded before any appreciable change in pressure occurs. Falls in pressure occur quite quickly, not so much as a result of outflow through natural or artificial channels, but rather as caused by expansion of the eyeball; though with this expansion, a steady state may be reached.

Experiments on eyes and on rubber models are described in an interesting paper by Phillips and Shaw [12]. Rubber models and the cornea show very similar stress–strain curves, "softening" with increasing stress; whereas tests on a scleral strip show the opposite effect. These authors also give useful information concerning different types of tonometer. Walker et al. [13] report interesting experiments using rubber diaphragms: also work on enucleated rabbits' eyes.

A fairly recent and very good review article was published by Macri and Brubaker [14]. Following a brief description of the chemistry of the aqueous humour and its processes of formation and flow, the authors describe in detail the principal ways of measuring the pressure.

With anaesthetized animals, a needle can be inserted into the eye and connected to a suitable manometer. In this way, continuous measurements of changing pressures can be studied. With human patients, as already indicated, rheometers are required that will record pressures without penetrating the eye. Macri and Brubaker give a historical account of the development of these instruments which can only be summarized here.

The indenter may be applied either to the cornea or to the sclera* and may measure either a single initial pressure ("static") or progressive changes in pressure (dynamic). The authors suggest that the static type of tonometer is more widely used than the dynamic, though the latter may be of importance for research purposes.

An interesting type of instrument from the rheologist's point of view is the ballistic tonometer. These instruments effect a collision of a known mass with the cornea. Some instruments measure the acceleration, some the impact duration, and some the rebound (see Vogelsang [15] and Dekking and Coster [16]). Another type of impact tonometer measures the duration of impact from the resonance frequency of a vibrating mass placed against the eye (see

---

* The cornea is the transparent membrane in front of the eye, through which the light passes. The sclera is a dense white fibrous coat that surrounds the eye, except where the optic nerve pierces it at the back and in the front, where it is modified to form the transparent cornea.

Roth and Blake [17]). Other ingenious methods, such as the measurement of the deflection of the cornea resulting from a standard pulsed air-jet, have been described.

At the time of writing, the authors believed that the most widely used tonometer was that of Goldmann [18] in which a circular contact of fixed diameter produces an intraocular volume charge of only 0.5 mm^3. There are two possible sources of error: the surface tension of the liquid "tear" and the effect of corneal bending; but these tend to cancel one another.

For books on tonometry see end of this chapter.

Vitreous humour

Not very much rheology would appear to have been done on the vitreous humour, with the exception of the researches of Pfeiffer [19–21] who also studied the optical properties. The latter work will not be discussed here.

The vitreous humour, from the eyes of mice and rats, drawn out into threads, shows many interesting optical and rheological properties. These include flow birefringence, rheodichroism, spinability and flow elasticity, all of which Pfeiffer measured. He found, among other things, that the Young's modulus of the threads increased rapidly at very low pH values, passing a sharp maximum at about pH 4.7, and then falling to a steady value around pH 6.0. However, no attempt was made to relate the results of rheological measurements to clinical conditions.

(Readers interested in more detailed work on the rheology of the eye should refer to the journal "Eye Research" over a period of some years.)

Other body secretions

(a) Prenatal mammary secretions

In 1938, Waller [22] drew attention to the frequency of engorgement, of mastitis and even of breast ulcers with the onset of lactation in women. He found a considerable improvement if the pre-natal secretion was removed manually, though some other medical specialists in the field regarded the improvements as more likely to have resulted from other factors. It was already known that removal of pre-natal secretion produces marked changes in the physical and chemical properties of the fluid in the case of the cow (see Woodman and Hammond [23] and Asdell [24]) and it was thought possible that similar effects might be found in human lactation. Hitherto, only qualitative estimates of viscosity appear to have been made.

Waller suggested to the present author that, if the very high viscosity "fluid" had a yield value, or showed a big increase in viscosity at low shear rates, there might arise mechanical difficulties in its passage through the nipple. Two investigations were therefore undertaken. In the first, Scott Blair [25] designed a very simple viscometer, of a type since used by many other workers for other materials. This consisted of a horizontal glass capillary attached at both ends to wider tubes which were bent upwards to a vertical position. One of these bore a scale, the zero position of which could be adjusted (see Chapter II). About 1 ml of secretion was put into this visco-meter and sucked (or blown) up to produce a convenient head. (The scale is previously set so that the zero corresponds to the position of the menisci when they are level in the two tubes.) The apparatus is held at 37°C in a glass-sided container and the height of the falling column, when pressure is released, is followed with a stop-watch. The full calculation, including a surface tension correction, is given in the original paper, though this correction can generally be ignored.

For a Newtonian fluid, a plot of time (t) vs. log (h_0/h), where h_0 is the initial head and h that at time t, would give a straight line passing through the origin.

Experiments showed enormous variations in viscosity but only a slight shear thinning, with no evidence of a yield value (see Fig. X-2). These results

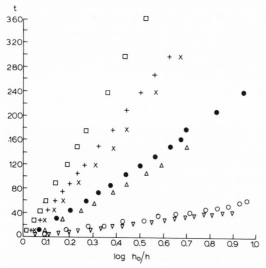

Fig. X (2). Flow curves for human prenatal mammary secretions. Reproduced from: Scott Blair, G.W., Biochem. J., 35: 267 (1941).

suggest that a blocking of the nipple caused by the consistency of the milk is not responsible for the engorgement.

Using this, and other (chemical) tests, Waller et al. [26] made a wider study of the whole question. They concluded that the removal of the antenatal fluid does not change its later characteristics, specifically the total nitrogen content, which is linearly related to the protecting power of the colloids, as measured by the "gold number". (The gold number is the weight of a protecting colloid just insufficient to prevent a change in colour of a standard red gold sol under specified conditions.)

As was to be expected, there was a striking fall in the globulin/casein ratio after parturition, but the truly "colostral period" was found to be very short: only a few days. Confirming Scott Blair's findings, there were enormous differences in the viscosity of the secretions. There was an approximate linearity between the viscosity and the logarithm of the concentration (Arrhenius' equation). The authors also agreed with Scott Blair that difficulties in withdrawal of the milk are due primarily to faults in the mechanism of expulsion rather than to rheological properties of the secretion.

The viscosity of the milk of other mammals has, of course, been measured; especially that of the cow. It was at one time hoped that viscosity or viscous anomalies might provide a quick method for determining chemical composition, especially in relation to the rather prevalent mastitis; but this is not so. In any case, the interest lay more in possible help to the dairy industry than to biology and would not fall within the scope of this book. A good review article, for those interested, was published by Cox et al. [27] in 1959, and little has been published since then.

(b) Lymph

The author can find very little in the way of rheological studies of lymph. Money et al. [28] measured the viscosity and density of endolymph and perilymph* from the inner ear of the pigeon, using a sphere of tungsten carbide rolling down a glass tube (0.030 cm diameter). Viscosities of 1.19 and 0.76 cp, respectively, were reported. "The generally accepted mechanism of the semi-circular canals is modified by the suggestion that different factors are important in the response to small and large stimuli" (quoted from Brit. Rheol. Abstr., No. 346 (1972)). Lymph has many characteristics similar to those of blood, but its rheological properties are more complex.

---

* The "membranous labyrinth" of the ear lies inside the "bony labyrinth" and is filled with a fluid called "endolymph". The fluid lying between it and the bony labyrinth is called "perilymph".

(c) Mucoproteins from urine

These have been studied by Tamm [29, 30] and his colleagues, using many measurements, including viscosity. They conclude that the molecular weight is approx. 7×10^6 (from centrifuge data). Diffusion experiments suggest that the molecules are long filaments. The viscosity is reduced by drastic heating.

(d) Mucus in general

Litt [31] has made an interesting comparison between the variations of the complex modulus with frequency (see Chapter II) for bronchial, cervical and aural mucus. The first two curves show a steep rise at low frequency, followed by a slight fall (a nearly constant portion), again followed, at high frequencies, by a steep rise. The curve for the ear mucus is sigmoid, showing a flatter portion in the middle range, but no minimum.

(e) The alimentary tract

In view of the importance of digestion and the prevalence of constipation, it seems surprising that so little rheology appears to have been published on the behaviour of the stomach and intestines and on their contents.

A recent article by Lew et al. [32] hardly borders on rheology. It is concerned with the transport phenomena in peristalsis; particularly in the small intestine. The authors quote a number of earlier papers dealing with the form of the sinusoidal motion of the tube wall. Since Lew et al. assume a model of the small intestine as a cylindrical tube with travelling nodal constructions and its contents (the chyle) as a Newtonian fluid, it is clear that this paper is not concerned with the rheological (as distinct from hydrodynamic) properties of the system, but readers interested in hydrodynamics should consult this paper and its bibliography.

In 1959, Heatley [33] described "some experiments on partially purified gastro-intestinal mucosubstance". Secretions from the duodenal pouch of the pig were filtered, after sterilization, through a sinter-glass filter. As Ogston and Stanier had found for separating hyaluronic acid from synovial fluid, this allows salts and small molecules to pass through, while retaining the mucus. This mucus has a gross structure that is destroyed by vigorous stirring, but not by repeated washing. It shows spinability and also the Weissenberg effect: i.e. it climbs up a rod rotated in it. When dilute mucus is mixed with phosphormolybdic acid in dilute hydrochloric acid, it forms short fibrous threads. On standing, these form a U-shape; presumably because the centre of the thread is heavier than its ends. Heatley also studied the solubility and the viscosity of the mucus in relation to pH and electrolyte concentration. A graph is given showing the effects of repeated shearing (in an Ostwald viscometer) and the even more dramatic fall in viscosity following high-speed stirring with a wire. Somewhat naturally, although good reproducibility was found for measurements of viscosity in the "beaten" samples, this was not so for the unbeaten samples. But the author quite rightly points out that the Ostwald viscometer is not the best instrument for this type of material.

In view of the importance of NaCl in cervical mucus (see Chapter VII) it is interesting that the viscosity* was not much affected by variations in NaCl concentration over a wide range: 0.01–1 M.

A kind of yield value was also measured: the pressure required to keep the mucus just moving in a standard capillary tube (internal diameter approx. 7 mm). A rough subjective measure of "stickiness" was also made. The mucus is not sticky, in the absence of electrolytes at pH values not above 4. These properties (which are fairly typical of mucus in general) are quite remarkable when it is appreciated that the concentration (as dry weight) is only about 0.1 %.

In 1969, Curt and Pringle [34] studied the viscosity of gastromucus in relation to duodenal ulceration. They also gave references to a number of earlier works all of which need not be listed here. But the following should, perhaps, be noted: Janowitz and Hollander [35] measured the viscosity of cell-free canine gastric mucus and Zalaru [36] also measured gastric mucus viscosity. But both these authors used capillary viscometers and Curt and Pringle rightly point out that the rate of shear is not constant under constant pressure in a capillary. However, as explained elsewhere in the present book (see Chapter II), this does not altogether rule out the use of capillary viscometers for studying non-Newtonian systems. Curt and Pringle used a Wells–Brookfield (coaxial cylinder) microviscometer, which is certainly better. They found that gastric mucus is both shear thinning and thixotropic; and, more important, that the viscosity, falling as the rate of shear rises from about 1–230 s^{-1}, is consistently higher for mucus from duodenal ulcer patients than that from controls, though the curves converge at the highest shear rates.

In a much later paper, Snary et al. [37], using pig gastric mucus, measured the viscosity in various concentrations of KCl. (The intrinsic viscosity does not appear to have been measured.) There are two sharp transitions, giving flat portions to the viscosity–KCl concentration curve, corresponding to 0.05 M and between 0.5 and 1.0 M KCl. Various conclusions are drawn concerning molecular configurations.

---

* We follow the author in using the term "viscosity" throughout, but of course the system is non-Newtonian: "consistency" would be more strictly correct.

Lykoudis et al. [38], referring to earlier work on peristalsis in the upper urinary tract (Lykoudis [39]), discuss some biorheological aspects of motility in the large intestine. The authors point out that there appears to be little or no published work on the contents of the colon or on human faeces. They proceed to show that the latter give a power equation relating shear rate to stress. Although there were (as might be expected) wide variations in the coefficient measuring the consistency, it is remarkable that the exponent remains always at about the fourth power of the stress. The authors conclude that "the theoretical results provide us with a correlation between the kinematics of the boundary of the colon, its geometry, the rheological properties of chyme [its contents] and the intraluminal pressure".

Miscellaneous

(f) Eggs

Quite a few papers have been published on the rheology of egg-white, but these are mostly concerned with the use of this material in foodstuffs and do not come within the scope of this book. A few articles are, however, worth quoting. As early as 1917, Stocks [40] found that egg-white consists of two distinct portions, one much thicker than the other. The flow properties of these two separated portions were studied, using a capillary viscometer, by St. John and Green [41]. The "thick" portion softens in the course of time if the egg is left standing open, but the "thin" white is not so affected. The consistency of the "thick" portion, especially, is obviously affected by the shearing in the viscometer; and by modern standards, the data cannot be said to have very much meaning.

In 1936, Wilcke [42] made some much more interesting measurements. The quality of a (raw) egg depends, to a considerable extent, on its consistency. Wilcke mounted the unbroken egg on a swinging torsion pendulum and measured the damping. The lower the consistency of the inside of the egg, the more energy is used up as a result of flow, and hence the greater the damping. Using glycerine–water mixtures as controls, it was shown that the damping with a Newtonian liquid passed through a maximum within a certain viscosity range. By controlling the moment of inertia of the system, it was possible to ensure that the consistency of the eggs gave damping below this maximum. It depended on the weight, but not on the eccentricity of the egg and differed characteristically for individual hens.

Brooks and Hale [43] measured the strength of hen's egg shells as well as their hardness. (The work was later summarized by Brooks [44].) The eggs were squeezed and crushed between two parallel plates. Load and deformation at rupture were recorded and related to the thickness of the shell. There was no correlation with the size of the egg, but some meaningful relation to its shape was found. The "Vickers hardness number" (penetration) was also measured and showed a high correlation with the strength. There was probably some correlation with the magnesium content of the shell.

Brooks, in his later paper, studied the displacement of nickel particles by a magnetic field in the two (thick and thin) portions of the white and described also a technique for measuring the bursting strength of the membrane. Muller [45] studied the Weissenberg ("rod climbing") effect on the two fractions of egg-white.

On the physiological aspects of egg development, McCafferty et al. [46] studied the embryonic development of the hen's egg but the methods used were not mainly rheological.

Very different are the eggs of the sea urchin. Hiramoto [47] has written a very comprehensive review article on their rheology, including over 130 references. These eggs are important in research on protoplasm (see Chapter III). Hiramoto discusses the structure of the egg. The pressure difference which exists between the interior (the endoplasm) and exterior of the cortex varies from 12 to 120 dynes/cm². Both the cortex and the endoplasm are viscoelastic, but the consistency is higher in the outer part of the cortex than in the endoplasm. The properties of the endoplasm may be approximated to those of a Kelvin model (a dashpot and spring in parallel) plus a dashpot in series. The properties change during fertilization and in early development in a very complex manner.

(g) Fibre formation in the silkworm and silk spinning

Not unnaturally, this subject has aroused most interest in Japan (Iizuka [48]). The thread produced by the domestic silkworm (*Bombyx mori* L) is known as "bave". It seems that the mechanical properties of the bave differ according to the race of the silkworm and no general relationships have been found. Fig. X-3 shows how the static and dynamic Young's modulus of the bave fall with increasing size. The dynamic modulus is seen to be higher than the static modulus. The degree of crystallinity was determined by X-rays and this was also found to be linearly related to the size, regardless of the race of silkworm and the selected portion of the cocoon. The principal constituents of the silk are fibroin and sericin. The visco-elastic properties are mainly determined by the former. Mechanical stresses applied during the extrusion from the silk gland produce coagulation of the silk fibroin.

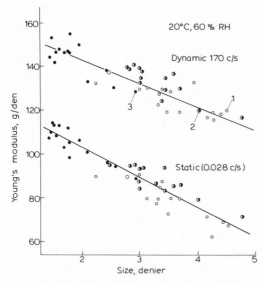

Fig. X (3). Young's modulus vs. size for bave. 1, outer portion; 2, middle portion; 3, inner portion. Reproduced from: Iizuka, F., Biorheology, 3: 3 (1965).

In his second paper, Iizuka describes measurements of the critical shear rate at which the molecule of silk fibroin denatures at different values of pH, ionic strength etc. The molecular chain is unfolded by stress to form the fibrous structure, stimulated by the presence of bivalent cations, such as Ca^{2+} and Mg^{2+}. The mechanisms of spinning are also discussed.

(h) Deoxyribonucleic acid (DNA)

Deoxyribonucleic acid (DNA) is of such topical interest that a study of the rheology of its solutions is important. Good reviews are given by Robins [49], including 75 references (see also an earlier paper by Butler and Robins [50]). Most of what follows here is a summary of his review.

Robins is concerned mainly with work published between 1961 and 1966, since the earlier work is described in an article by Yang [51]. The present brief summary will not include references to the individual papers, for which the reader should consult Robins' review.

The study of the structure of the DNA molecule has involved many different techniques as well as those of rheology and is linked with the vast field of work concerned with the structure of macromolecules in general.

Although, in theory, a flexible chain molecule with a random structure should show Newtonian behaviour, in practice, almost all high polymers are non-Newtonian. This is due to various factors: an internal viscosity arising from a limited rotation of bonds during molecular deformation, and the anisotropy of hydrodynamic interaction within the macromolecule. There is also what is known as the "excluded volume factor": if the monomer–solvent associations are strong (in what is called a "good solvent") the expanded molecule becomes deformed by shear forces and the system is non-Newtonian.

In general, an equation of the following form is used:

$$[\eta]\dot{\gamma} = [\eta]_0 (1 - a\,\dot{\gamma}^2 + b\,\dot{\gamma}^4 + \ldots)$$

(present author's symbols)

when $[\eta]\dot{\gamma}$ and $[\eta]_0$ are the intrinsic viscosities at shear rates $\dot{\gamma}$ and zero, respectively, a and b are rather complex constants.

Even powers of $\dot{\gamma}$ are used "because the decrease in with increasing $\dot{\gamma}$ must be independent of the numerical sign of $\dot{\gamma}$". (The present author has shown the fallacy of this argument, which has been applied to many rheological equations. It depends on a confusion of two different meanings of the term "negative": see Scott Blair [52].)

However, in practice, the equation, which is very general, works quite well, especially for the case of a rigid rod or ellipsoid in revolution. However, DNA is too flexible for such a treatment, but not flexible enough to be treated as a simple polyelectrolyte. A "stiffened coil" (worm-like) model has recently been proposed, but it is difficult to apply any of the treatments commonly associated with high polymers to DNA, since its molecules interact so strongly with their neighbours. The most effective work has been done at very low shear rates, part of the difficulty being that the molecule is very readily broken down by shearing. The highest intrinsic viscosity reported before 1966 is about 3000 dl/g, which suggests a molecular weight of about 2.5×10^9–15×10^9. Solutions are still non-Newtonian at shear rates as low as 0.1 s^{-1}.

All the usual types of viscometer: capillary, coaxial cylinder, etc., have been used in these investigations.

Robins [49] discusses at some length the probable mechanism of the shear breakdown of DNA solutions. It seems likely that there is a real breaking of the molecular backbone, which is not recovered when the DNA is precipitated and redissolved.

As a further complication, it seems that the differences that undoubtedly exist between DNA solutions from different sources are too subtle to be detected rheologically. With extreme care, however, the whole DNA content of a bacteriophage can now be isolated as a single particle. A balance has to be found between a maximal degree of purity and minimal handling.

Joly [53] has contributed an interesting article on the flow birefringence of different concentrations of DNA. He has isolated fractions which differ in their optical properties in dilute solutions. The degree of anisotropy increases with increase in the size of the particles, yet these same samples, within the same range of concentration, show no anomalies in viscosity, nor on centrifuging. This anomaly is difficult to explain.

(i) Locusts

Although at the time of writing (November 1972), the work of Vincent and Wood [54] is still at an early stage, some account of their published results should be included in this "miscellaneous" section. The adult female locust (*Locusto migratoria*) digs a hole, 8–9 cm deep, in the sand in order to deposit her eggs. The normal length of the ovipositing part of the abdomen is about 2.5 cm. Although part of the increase in length needed to deposit the eggs is produced by unfolding of membranes, there must be an extension of a part of the membrane of the order of 15-fold. If there is no sand, the unfolding process is not followed by any elastic extension. Preliminary experiments have shown that the membrane is visco-elastic and the nature of the elasticity is the subject of continued studies.

(j) Dragon-fly wings

The nearest approach to be found in the animal kingdom to natural rubber, is a material that can be extracted from the wings of dragon-flies. This substance has been studied by Weis-Fogh [55] and named "resilin". It is excreted in thick continuous layers by the epidermis. When swollen in water, it can be compressed to 1/3 or extended to 3 times its initial length, and has much the same elastic modulus as rubber. Its elasticity is evidently of the entropy type. The same author has shown (Weis-Fogh and Anderson [56]) that the "elastin" obtained from bovine ligaments differs fundamentally from rubber in its elastic behaviour.

(k) Surface rheology

Much valuable information can be obtained by studying the stress–strain relations of monomolecular films, spread on the surface of a liquid. These films can behave as two-dimensional gases (molecules independent of one another), or as liquids or as solids. Most of the work that bears on biorheology has been concerned with proteins and does not itself really come within the scope of the present book.

The leading worker in this field, M. Joly [57], has published a very full review of work on this subject, to which the interested reader is referred.

(l) The bladder

Mellanby and Pratt [58] studied long ago the rather complex properties of the bladder, working on an anaesthetized cat. The bladder tends to show changes in volume characteristic of the internal pressure, the higher the pressure, the larger the volume. But there would seem to be no simple relation between pressure and volume.

At a constant pressure, there are rhythmic changes in volume. The effects of anaesthetics, adrenaline, acetylcholine and other substances were studied. The authors quote Elliott as describing the bladder as a thin-walled hollow elastic sphere in which the tension is roughly proportional to the volume, multiplied by the hydrostatic pressure.

REFERENCES

[1]   Bowman, W., *Lectures on the Parts Concerned in the Operations on the Eye and on the Structure of the Retina* (Longman, London, 1849).

[2]   Eisler, P., *Albrecht v. Graefes Arch. Opththal.*, 124: 705 (1930).

[3]   Fisher, R. F., *J. Physiol.* (Lond.), 201: 1 and 21 (1969); 213: 147 (1971); 228: 765 (1973).

[4]   Finsham, E. F., *Brit. J. Ophthal.*, 21: Monogr. Suppl. 8 (1937).

[5]   Vogt, A., *Lehrbuch und Atlas der Spaltlampenmikroskopie des lebenden Auges*, Vol. 1 (Springer-Verlag, Berlin, 1930).

[6]   Fukada, M., *Jap. J. Ophthal.*, 7: 47 (1963).

[7]   François, J., Rabacy, M. and Wieme, R. J., *Arch. Ophthal. N. Y.*, 53: 481 (1955).

[8]   Fisher, R. F., *Trans. Ophthal. Soc. U.K.*, 90: 93 (1970).

[9]   Woo, S. L-Y., Koboyashi, A. S., Schlegel, W. A. and Lawrence, C., *Europ. Eye Res.*, 14: 29 (1972).

[10] Amsler, M. and Huber, A., in *Deformation and Flow of Biological Systems* (Ed. A. Frey-Wyssling), p. 149 (North-Holland Publ. Co., Amsterdam, 1952).

[11] Friedenwald, J. S. and Moses, R., *Docum. Ophthal.*, 4: 335 (1950).

[12] Phillips, C. I. and Shaw, T. L., *Exp. Eye Res.*, 10: 161 (1970).

[13] Walker, R. E., Litovitz, T. L. and Langham, M. E., *Exp. Eye Res.*, 13: 14 (1972) and 13: 187 (1972).

[14] Macri, F. J. and Brubaker, R. F., *Biorheology*, 6: 37 (1969).

[15] Vogelsang, K., *Ber. dtsch. ophthal. Ges.*, 48: 106 (1930).

[16] Dekking, H. M. and Coster, M. D., *Ophthalmologica*, 154: 59 (1967).

[17] Roth, W. and Blake, D. G., *J. Amer. ophthal. Ass.*, 34: 971 (1963).

[18] Goldmann, H., *Bull. Soc. Ophthal. France*, 67: 474 (1955).

[19] Pfeiffer, H. H., *Naturwissenschaften*, 11: 398 (1963).

[20] Pfeiffer, H. H., *Biorheology*, 1: 111 (1963).

[21] Pfeiffer, H. H., in *Symposium on Biorheology, Proc. 4th int. Congr. Rheol.*, Providence, R. I., 1963 (Ed. A. L. Copley), Part IV, p. 535 (Interscience, New York, 1965).

[22] Waller, H., *Clinical Studies in Lactation* (Heinemann, London, 1938).

[23] Woodman, H. E. and Hammond, J., *J. agric. Sci.*, 12: 97 (1922) and 13: 180 (1923).

[24] Asdell, S. A., *J. agric. Sci.*, 15: 358 (1925).

[25] Scott Blair, G. W., *Biochem. J.*, 35: 267 (1941).

[26] Waller, H., Aschaffenburg, R. and Grant, M. W., *Biochem. J.*, 35: 272 (1941).

[27] Cox, C. P., Hosking, Z. D. and Posener, L. N., *J. Dairy Res.*, 26: 182 (1959).

[28] Money, K. E., Bonen, L., Beatty, J. D., Kuehn, L. A., Sokoloff, M. and Weaver, R. S., *Amer. J. Physiol.*, 220: 140 (1971).

[29] Tamm, I., Bugher, J. C. and Horsefall, F. L., *J. biol. Chem.*, 212: 125 and 135 (1955).

[30] Porter, K. R. and Tamm, I., *J. biol. Chem.*, 212: 135 (1955).

[31] Litt, M., *Ann. Otol. Rhinol. Laryng.* (St. Louis), 80: 330 (1971).

[32] Lew, H. S., Fung, Y. C. and Lowenstein, C. B., *J. Biomech.*, 4: 297 (1971).

[33] Heatley, N. G., *Gastroenterology*, 37: 304 (1959).

[34] Curt, J. R. N. and Pringle, R., *J. Brit. Soc. Gastroent.*, 10: 931 (1969).

[35] Janowitz, H. D. and Hollander, F., *Gastroenterology*, 26: 582 (1954).

[36] Zalaru, M. C., *Med. interna. Buc.*, 18: 89 (1966) (in Rumanian).

[37] Snary, D., Allen, A. and Pain, R. H., *Europ. J. Biochem.*, 24: 183 (1871).

[38] Lykoudis, P. S., Patel, P. D. and Picologlou, P., *Biorheology*, 9: 158 (1972) (Abstr. from *Proc. 6th int. Congr. Biorheol.*).

[39] Lykoudis, P. S., in *Urodynamics* (Ed. S. Bayarsky) (Academic Press, New York, 1971).

[40] Stocks, H. B., *First Report of Colloid Chemistry* (British Association for the Advancement of Science, London, 1917).

[41] St. John, J. L. and Green, E. L., *J. Rheol.*, 1: 484 (1930).

[42] Wilcke, H. L., *Iowa State Coll. Res. Bull.*, p. 194 (1936).

[43] Brooks, J. and Hale, H. P., *Nature*, 175: 848 (1955).

[44] Brooks, J., in *Texture of Food, Soc. Chem. Ind. Monogr.* (Lond.), No. 7: 149 (1960).

[45] Muller, H. G., *Nature*, 189: 214 (1961).

[46] McCafferty, R. E., Pressman, S. H. and Kniseley, W. H., *Biorheology*, 2: 171 (1965).

[47] Hiramoto, Y., *Biorheology*, 6: 201 (1970).

[48] Iizuka, E., *Biorheology*, 3: 1 (1965), and 3: 141 (1966).

[49] Robins, A. B., *Biorheology*, 3: 153 (1966).

[50] Butler, J. A. V. and Robins, A. B., in *Flow Properties of Blood and Other Biological Systems* (Eds. A. L. Copley and G. Stainsby), p. 337 (Pergamon Press, Oxford, 1960).

[51] Yang, J. T., *Advanc. Protein Chem.*, 16: 232 (1961).

[52] Scott Blair, G. W., *Rheol. Acta*, 11: 238 (1972).

[53] Joly, M., *J. Polymer Sci.*, 29: 77 (1958).

[54] Vincent, J. F. V. and Wood, S. D. E., *Nature*, 235: 167 (1972).

[55] Weis-Fogh, T., *J. molec. Biol.*, 3: 520 (1961) and 3: 648 (1961).

[56]   Weis-Fogh, T. and Anderson, S. O., *Nature*, 227: 718 (1970).
[57]   Joly, M., *Surface Colloid Sci.*, 5: 1 (1972) (in two parts).
[58]   Mellanby, J. and Pratt, C. L. G., *Proc. roy. Soc. B*, 127: 307 (1939).

BOOKS ON TONOMETRY

Draeger, J., *Tonometry* (Hafner, New York, 1966).
Gloster, J., *Tonometry and Tonography* (Churchill, London, 1966).

Chapter XI

PSYCHORHEOLOGY

The validity of measurements of sensations

There is much overlapping between psychology and a number of other sciences. Psychorheology has been mainly concerned with relating the results of subjective assessment of the consistency of industrial materials with the data from rheological tests. As such, it is not generally connected with biology.

However, when the baker judges the consistency of his dough, or the potter of his clay by the "feel" in his hands, the sensations from which he forms his judgements are derived from sources which are cutaneous (surface of the skin) and kinaesthetic (movement of the joints). These are biological phenomena.

For this reason, we shall include in this book some account of the history of the rather broader field of "psychophysics" as a whole, as well as some specific applications to rheology.

The classical work of Weber (1834) is well-known. He found *experimentally* that the "just noticeable differences" (j.n.d.; ΔE) in the intensity of a stimulus —say, a beam of light is proportional to the intensity of the stimulus. Later research has shown that this is approximately correct in many cases. However, in 1850 (though apparently not published until 1860) Fechner "extended" this law by making certain assumptions, which were hotly contested, especially by Tannery (1875) and by Bergson [1] in 1889.

Fechner extended Weber's concepts in several ways: First, he assumed that our consciousness of an increase in stimulus is produced by a corresponding increase in sensation (ΔS). He further assumed that equal increases in j.n.d. correspond to equal increments of sensation, or:

$$\Delta S = C \frac{\Delta E}{f(E)}$$

where C is a constant and the function f is replaced by E when Weber's law holds. He then changed from small finite differences (Δ) to differentials and, integrating, he wrote $S = C \ln (E/Q)$ when Q is a constant. Bergson challenged each one of these assumptions. Without going into his detailed criticisms, they may be summarized by saying that he maintained that it is not possible to "measure" sensations.

Bergson believed that sensations are not purely quantitative but also qualitative and that only quantities can be "measured". For example, if I put my hand into water at just above the freezing point, I have a definite sensation of pain. As the temperature is progressively raised, this sensation changes to one of pleasure (or at least comfort) around 40°C, but it again becomes pain when about 60°C or more is reached. The gradually changing temperature is a "quantity" and can be continuously measured on a thermometer scale and water at two different temperatures can be simultaneously present "in space", whereas sensations cannot. (This last statement is arguable, as in the famous experiment when a subject puts one hand in rather warm water and the other hand in rather cold water. When he puts both hands into tepid water, the same water will feel cold to one hand and warm to the other.)

Meanwhile, Delboeuf, in Belgium, did experiments which, he claimed, made it quite possible to persuade subjects to judge a stimulus as being twice as great as a given stimulus, or to halve the interval between stimuli, thus making it possible to test Fechner's equation. Although he found it necessary to modify the equation somewhat, he claimed to show that Bergson was wrong and that it was perfectly possible to make a linear scale of sensations. Bergson's reply does not, today, seem very convincing: he wrote that Delboeuf's "psychophysics" is merely "a symbolic interpretation of quality as quantity, a more or less rough estimate of the number of sensations which can come in between two given sensations". But this is surely a "measurement"!

Probably the real difference of viewpoint lay in there being no very clear definition of "measurement". Much later, this point was taken up by Stevens, who published many papers on the subject, of which only two will be quoted here (Stevens [2, 3]).

Stevens defined "measurement" as "the assessment of numerals to things so as to represent facts and conventions about them". He originally proposed four types of measurement: nominal, ordinal, interval and ratio. Later, he subdivided the interval group into "linear" and "logarithmic" but there is considerable confusion in his use of the term "logarithmic". In his writings it is often not clear whether he means "exponential" (e.g. $y = e^x$) or a power equation (e.g. $y = x^a$).

Normal "measurements" would hardly qualify as measurements, except within Stevens' definition: e.g. the arbitrary numbering of soldiers or prisoners. "Ordinal" scales place the members in a meaningful order, e.g.

soldiers in line, tallest on the right, shortest on the left, but no equality of interval between them. "Interval" scales have equality of intervals but arbitrary zeros: i.e. they cannot be interconverted by a simple multiplication, e.g. Fahrenheit and centigrade temperatures. Finally, the highest category of physical measurements are called "Ratio": e.g. mass, length, absolute temperature, etc.

Stevens pointed out the dangers of using any but the simplest statistical methods for the study of data other than those on a ratio scale, even though psychologists and many others often use forms of multivariate analysis involving correlation coefficients for "ordinal" data. In so far as this works, it depends on an unprovable approximate equality between the intervals.

It is interesting that Stevens, in his turn, was subjected to criticisms which strongly recall the Delboeuf–Bergson controversy.

In quite early times, alternatives to the Fechner equation were proposed. Plateau, and perhaps still earlier, Brentano, had suggested a power equation to replace Fechner's exponential. Stevens [4, 5] proposed that there are two types of stimuli, one of which leads to Fechner's equation and the other to Plateau's. Those stimuli concerned with answering the question "how much?", e.g. measuring a threshold for sound intensity, he called "prothetic"; and those in which the intensity of the stimulus does not change and where the question is "of what kind?", e.g. the pitch of sound, he called "metathetic". The former, he claims, will follow Fechner's equation and the latter, Plateau's.

Treisman [6, 7] criticized this distinction, claiming that the form of the equation obtained would depend simply on how we define sensation and, if the present writer understands rightly, that we learn to make ratio comparisons for such modes as sound intensity but not for pitch. Different psychological tasks show different measures of stimulus intensity.

However, in spite of Prof. Treisman's kindness in having personal discussions with the author, the latter does not feel that he really understands Treisman's position: it seems to him that the controversy between Stevens and Treisman is in many ways similar (though not identical) with that between Delboeuf and Bergson some eighty years earlier.

Meanwhile, "whether sensations are truly measurable or not, we can certainly record what they say they feel" [8]*. (For an excellent essay on the

---

* In this paper, the author has quoted a little rhyme of which he cannot trace the origin:
"There once was a craftsman of Deal
Who judged all his products by 'feel'
But he spent his brief leisure
In trying to measure
Sensations he knew were not real."

psychology of craftsmanship, the reader is referred to two quite early papers by Harper [9].) Dr. Harper is both a psychologist and a physicist and other aspects of his work will be quoted later in this chapter.

Following this brief introduction to the general status of psychophysics, we can now consider some of its specifically biological aspects.

Subjective assessments*

The psychological basis of the assessment of rheological properties, generally in the hands or mouth, was fully discussed in an earlier book by the present author [10], but since this book is now long since out of print, some modernized recapitulation will be given here.

The original treatment was based on the principles of Gestalt Psychology, now somewhat "*démodé*", nevertheless it is undisputed that our subjective appreciation of rheological properties must depend on a fusion of various sense perceptions giving a basis for judgement that is certainly more than a mere sum of its components. Some of the properties assessed lie on the borderline of rheology but a brief description of the experiments will not be out of place here.

An early pioneer in this field was Miss Sullivan [11, 12]. The finger of a blindfolded subject was dipped into various more or less liquid substances to study what Sullivan called "perceptions of liquidity, semi-liquidity and solidity". "The large class of experiences included in the term 'liquidity', are true blends or fusions, but the equally large class of experiences included in the term 'solidity' such as dry, soft, hard, etc., are not really touch blends at all, if we mean by touch blends fusions or integrations. They are rather cutaneous patterns or mosaics." (The present author is not at all sure that he agrees.) Nevertheless, Sullivan did a striking experiment in which the fingers were dipped into what (unknown to the subject) was, in fact, water at a series of temperatures. The sensations were described as follows:

Temp. 0°C   Mushy, like partially melted snow**
      10°C   Muddy, like mercury (*sic!*)
      15°C   Gelatinous, like gelatin
      25°C   Wet—water
      38°C   Oily, like oil
      42°C   Greasy, like melted butter.

---

* Some of the researches described in this section did include comparisons with measurements of viscosity, but those concerned with the interrelationship between subjective assessments and complex rheological properties are discussed in the following section.
** Were there perhaps ice crystals present?

In the third paper, solid materials were used: metals, wood, cotton-wool and velvet. "Hardness" was conditioned by definite boundaries to the specimens. There was a tendency to judge "soft" when samples were warm and "hard" when they were cold.

Experiments were also done blowing jets of hot or cold air onto the hot and cold spots on the skin. Solid cylinders of different weights (? densities) were compared, but the weight had no regular effect on the judgements of "hard" and "soft". The cylinders were rested on the hands and lateral movements of the limbs were not necessary for the perception of "hard" and "soft". In general, Sullivan concluded that it is very difficult to perceive a cold specimen as "soft", but sometimes possible to perceive a warm specimen as "hard", only, however, when the specimen had sharp boundaries.

Two other early workers, Zigler [13] and Meenes and Zigler [14], studied stickiness, roughness and smoothness. "Smoothness is a simple tactual perception constituted of a field of cutaneous pressure qualities, all of which possess a low or moderate uniform intensity, and a relatively uniform clearness, and are so clearly aggregated and blended as to give rise to an impression of extreme compactness. Roughness is a tactual perception constituted of an areal pattern, including both cutaneous and subcutaneous pressure elements. This perception is characterized by the lack of uniformity of stimulation. . . . Neither roughness nor smoothness is experienced when movement is lacking. In the absence of movement, the experiences are labelled evenness or unevenness."

Various physical measurements of the roughness of materials have been proposed, but little if anything appears to have been done to correlate them with subjective assessments.

Binns [15] investigated the alleged superiority of wool-top testers to unskilled subjects in assessing the softness of wool, claimed to depend on "feel" on the thumb. In fact, this superiority disappeared if the tests were done blindfold. Sight, and not "feel" was the operative sense. This result and much else was established in a much later study by Harper et al. [16]. An attempt was made to assess the interval between the successive quality numbers on a subjective "ordinal" scale.

Katz and Stephenson [17] compared the judgement of "weight" in pulling on an elastic spring with that when an ordinary load is lifted. It seems that the sensation of "heaviness" is less when, in fact, the same force is exerted on a spring than it is with a direct load.

A series of researches, over a long period of time, have been devoted to the assessment of viscosity of liquids in the mouth and also by squeezing between the finger and thumb. For an account of the early work see Katz

[18] and, for later researches, Renner [19]. Much later, Stevens and Guirao [20] asked subjects to assign numbers to represent the viscosity of silicone fluids (viscosities from about 0.1 to 1000 p) by shaking and turning a bottle containing the liquid, and by stirring the liquid with vision and also blind-folded. These authors found a linear relationship between the logarithm of the assessed viscosity and the logarithm of the measured viscosity. If subjects were asked to score "fluidities" (i.e. the inverse of viscosity), the double-logarithmic relation was not so good.

Fryklöf [21] (his paper is in Swedish but has a long English summary) using various ointments and creams found, unlike Stephens, an exponential relation between the subjectively assessed consistency and the viscosity for Newtonian liquids. In visual tests, the threshold varied widely over the range of viscosity used. In stirring experiments, the correlation between subjective and objective tests was twice as good when the subjects could see the samples. Fryklöf also studied the effect of increasing the time interval between com-parisons of samples with a standard (for non-Newtonian materials especially).

Yoshida [22], following up the work of Katz [18], made a very thorough study of the sense of touch, using the statistical method of multiple factor analysis. Four factors were isolated*: (1) heaviness–coldness; (2) smooth-ness–wetness; (3) hardness; and (4) a factor concerned with differentiating tactual and visual impressions.

Comparisons with complex rheological properties

Prof. David Katz, a distinguished psychologist, was obliged to leave Ger-many in the mid-1930's and, before taking up a Chair in Stockholm, it was the author's privilege to work with him in studying relationships between the extremely complex rheological properties of flour doughs and the baker's assessment of "body" which he gets by handling the dough. The rheological work will be only briefly mentioned here, since it is not in itself concerned with biology; but the correlations with subjective assessments made by handling come within the range of this book.

Katz [23, 24] published two papers on this subject in which he drew the following general conclusions. The baker does not "analyse" his sensations into elastic and viscous (or plastic) components as does the rheologist. He "gets a total impression of a general and somewhat vague character although he can analyse this total impression within certain limits, ill-defined, by changing his ways of handling the dough . . . but innumerable earlier experi-

*This reminds one very much of early Jain philosophy in India, though of course the latter had no statistical basis.

ences of similar and other doughs come into play and contribute to his final judgement".

Katz persuaded the baker to make judgements blindfolded, and also, he himself kneaded the dough, the baker judging its properties only visually. It seems that both touch and sight contribute to the baker's judgements. With regard to handling, Katz concluded that "springiness" (elasticity) played a more important part than did differences in "viscosity". Like Sullivan, he found that temperature greatly influenced judgements and that "stickiness" was more important than had previously been appreciated.

Studying both English and German bakers, Katz found that the amount of a "weak" flour (i.e. one that would make a small, poorly textured bread) that had to be added to a standard "strong" flour which could be detected by the bakers, was much greater than had previously been anticipated. This means that the bakers (even those highly skilled) could not detect differences in the "body" of the dough that would be quite easily recognized by rheological tests.

Rheologists sometimes invent an instrument that imitates the processes to which some material is subjected in manufacture or use and, because the instrument gives reproducible and apparently meaningful data, they assume that these must be related to what the craftsman or expert feels with his hands and associates with good or poor quality. An example of this was published by Scott Blair and Potel [25], who were comparing the rheological properties of doughs as used by bakers in England and in France. In the latter country, as a criterion of quality, an instrument was (and still is) used in which a "bubble" is blown in a disk of dough, the volumes and pressures being recorded. This imitates the expansion of the dough under the influence of the CO_2 produced by the yeast, and the reproducibility is good. However, data were available giving bakers' and millers' judgements of the "strength" of quite a large number of doughs and it was found that these assessments were not significantly correlated with the instrumental data.

In the case of flour doughs, the fundamental rheological properties—viscosities, elastic moduli and relaxation times—have been shown to be related to baking quality; though it is perhaps a pity that many of the excellent researches that have been done in measuring the rheological properties have been quite unconcerned with their relation to subjective expert assessments.

Marriott [26] showed that, in the pharmaceutical and cosmetic industries, there were reasonably good correlations between rheological data and such subjective assessments of quality as are made by touching with the finger or squeezing a tube (e.g. for shaving cream and toothpaste).

Prentice [27], having measured the consistency of butter using many instru-

154

ments, found a very high correlation with subjectively assessed "spreadability" and the force–flow rate relation when the butter is pushed through a metal tube. It is strange that an instrument that imitated the "spreading" fairly accurately did not show quite such good correlations. (Much of Prentice's work was also published in Research Reports of the British Food Manufacturing Industries Research Association around 1950.)

Working with fats, Uzzan and Sambuc [28] and Naudet and Sambuc [29] made a very thorough study of the relation between such properties as hardness, "spreadability" ("facilité d'étalement*) etc., of margarine, finding good correlations between the rheological data and the subjectively assessed properties.

Matz [30] in his book, "Food Texture", discussed at considerable length the psychophysical aspects of judgements of food quality in relation to rheological testing, and very recently Sone [31], in Japan, has discussed similar problems at a rather more academic level.

Since the comparatively new "Journal of Texture Studies" first appeared, this journal has published a number of papers which lie on the borderline of biorheology. Two papers especially should be noted: Parkinson and Sherman [32] point out that low viscosity Newtonian liquids exhibit turbulence in a cone-plate viscometer at high shear rates which are comparable to those in the mouth when subjects are assessing viscosity, though saliva may reduce this effect to some extent. Such materials appear to have a higher viscosity as assessed in the mouth than that measured in the viscometer; whereas with non-Newtonian materials, such as ice-cream, there is no turbulence and the presence of saliva produces a reduction in consistency in the mouth.

The other paper of note is a review article by Moskowitz et al. [33]. These authors quote a number of articles on psychophysics, not specifically concerned with rheology and therefore not given in this book.

It is clear that the various terms used to describe the rheological behaviour of materials fall into two classes. Scott Blair [34] has called these "connotative" and "denotative". These terms are used in rather a different sense from their original use by J. S. Mill and by J. M. Keynes: their "modernized" significance was borrowed by the author from Eysenck [35].

In rheology, as indeed in all branches of physics, we need certain precise terms to describe the behaviour of materials defined in (usually**) whole-number units of such fundamental concepts as Mass, Length and Time (MLT), or Force, Length, Time. (The number need not necessarily be three.)

---

* Other French authors translate this as "tartinabilité"!
** Some electrical quantities have powers of 1/2: see also below.

The definition of these terms can be changed by general agreement, as has been done in the case of "thixotropy". Some authors have attempted to give this term a rigid dimensional definition in terms of MLT and the original definition given by the inventor of the term (H. Freundlich) which limited the phenomenon to a slowly recoverable breakdown from a solid to a liquid state as a result of shearing, was extended, with Freundlich's approval, to include all falls in consistency followed by an *observably* slow recovery (see Chapter I). The "connotation" of the term was reduced but its "denotation" was increased. Very roughly, one might say that its connotation depends on its definition in dimensional terms, whereas its denotation depends on the range of materials which come within the definition. Naturally, as the number of materials studied is increased, denotation can be increased without loss of connotation.

Rheology, though quantitatively based on connotative concepts, cannot do without denotative terms. As an example, at one time an attempt was made to define "consistency" in MLT units; but, if this were done, we should need another term to describe "the assessable property of a material by which it resists permanent change of shape, defined by the complete stress–flow relation". Articles such as those of Moskowitz et al. [33] are full of very necessary denotative terms: e.g. sliminess, firmness, unctuousness (oiliness), tack, mouth-feel, etc. The aim of the rheologist is to attempt to analyse such terms, wherever possible, into connotative components, each of which will have, of course, less denotation than the original terms.

Moskowitz et al. [33] also discuss the statistical aspects of the handling of such data, on much the same lines as were followed by Stevens (see ref. 3).

Scott Blair's paper [34] is immediately followed by a valuable chapter by Harper on the psychological aspects of the problem. Harper has also published many other papers on similar lines, of which only two will be quoted here [36, 37]. Dr. Alina S. Szczesniak [38] (who is Co-Editor-in-Chief, with Dr. P. Sherman of the "Journal of Texture Studies") has published many valuable papers on "the correlation between objective and sensory texture measurements". Since these come on the boundaries of biorheology, only one general article will be quoted here.

Several papers have been published on the effects of pressure on the human body in assessing comfort in seating, especially in cars. The questions asked by Stone [39] are (1) whether it is possible to get agreement among people on the subjective preference for different types of cushioning and, (2) if agreement can be reached, can these preferences be related to the rheological properties of the cushions?

The twenty subjects (of whom nineteen were males) were asked which of pairs of seats they preferred "with regard to the *comfort of the cushion**, taking the cushion compressibility only into account". This would seem to be a somewhat restricted question. Subjects were also asked to place the eight seats in a rank order of "comfort". A second test, eight months later, on as many of the original subjects as were then available, was also done. The agreement between all these tests was satisfactory.

The situation was complicated by the fact that the rheological properties of the "foam" used in the seats were highly complex. For objective measurements a dynamic indentation recorder was used, a maximum load of 200 lb on a 305-mm-diameter disk was applied in continuous cycles. Load–deflection curves are shown and various arbitrary parameters were taken. Quite good correlations were found between selected parameters and judgements of comfort.

It is concluded, perhaps surprisingly, that most subjects prefer harder to softer cushions. The author suggests that this is because "the body weight should be supported on the haunch bones. The tissues over this area seem to be especially well adapted to withstand pressure of seating, whereas the thigh muscles are not meant for compression which is more likely to occur with soft cushioning." The author distinguishes between the psychological satisfaction in being enveloped in soft upholstery and the true comfort associated with a certain firmness. Cushioning with an increasing resistance to compression is more acceptable than is a material that decreases in resistance on immediate compression.

Rheology of mastication

In the food industry, the toughness, brittleness and many other "denotative" concepts, are of practical importance. It is often difficult to relate these to the standard rheological properties and indeed sometimes, as in the case of meat, it is quite difficult to measure anything meaningful: "lumps" of meat are far from homogeneous and individual fibres may not be typical.

For this reason, instruments have been designed to imitate as closely as possible the action of the human teeth and jaws. Pairs of dentures, attached to automatic "jaws" are made to grind the food in a standard manner, and the forces engendered are measured. Many subjective tests on panels of young subjects (with no dentures) have also been done and more sophisticated instruments exist. (The reader is referred to the "Journal of Texture Studies".) One very unusual method is, however, worthy of mention. Drake [40, 41] has devised an apparatus which, when attached to a hearing-aid earphone, magni-

---

* Italics in the original.

fies the sounds produced when food is chewed. Although it is amusing to hear, relayed on a loudspeaker, the sounds made by people chewing lettuce, biscuits or meat, the method gives quite seriously useful information in the food industry. In the second of the papers quoted, time frequency analysis diagrams are shown for bread, biscuits and raw carrot (Fig. XI-1).

Drake points out the potentialities of this method in the fields of orthodontics and surgery of the jaw joints. It is in such applications that the work would bear most directly on biorheology.

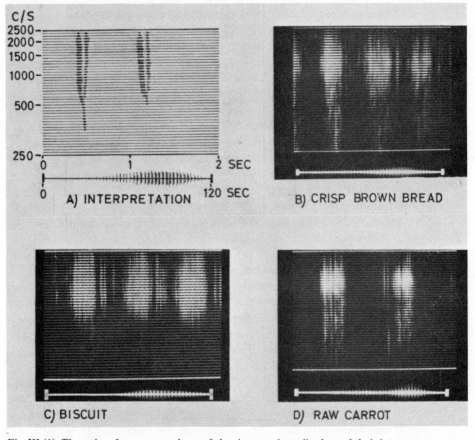

Fig. XI (1). Three time-frequency analyses of chewing sound amplitudes and their interpretation. Upper parts: time plotted horizontally and frequency vertically, with sound amplitude visualized as light intensity. Lower parts: time horizontally, sound amplitude vertically. Reproduced from: Drake, B., Biorheology, 3: 24 (1965).

Drake has also published work on relating such mechanical tests to subjective assessments; but much of this is written in Swedish, without summaries in other languages. However, he gave a general summary of the work, in English, at a Congress in Washington, D.C., in 1970 [42].

Analysis of pressure sensations in terms of rheological properties

Initially connected with a study of the assessment of the consistency of dairy products, principally cheese, by experts, a study of an essentially academic kind was made some thirty years ago, during a period of about ten years, on judgements of firmness of rheologically simple and complex materials by squeezing between the finger and thumb. These experiments were described in many articles at the time and are fully summarized in the author's earlier book, now out of print [10]. A brief résumé of the work will be given here.

It seemed that the first information needed was to find the threshold for distinguishing differences in elasticity of Hookean systems and in viscosity of very highly viscous Newtonian fluids. Steel springs, fitted with hard plastic disks at their ends, were incased in a soft tissue. These cylinders were all of the same size but differed in their compression moduli ("stiffness"). (In earlier experiments, rubber cylinders were used but, although these are approximately Hookean, they change their properties too quickly by deterioration.)

Californian bitumen was shown, by rheological tests, to be adequately Newtonian and could be diluted to give a suitable range of viscosities by the addition of an oil (see Scott Blair and Coppen [43]).

For elasticity it was found that an 80% level of correct judgements corresponded to about 9% difference in stiffness; and for viscous materials, to about 30% differences in viscosity.

In later work, comparisons for sensitivity were made between various categories of subjects, e.g. male–female, qualified and unqualified laboratory workers, age groups, skilled craftsmen, etc. (see Scott Blair and Coppen [44]). Only one group stood out as significantly different from the rest: routine analysts were found to be definitely superior in acuity.

This fact, combined with subjective observations of individual subjects, led to the conclusion that acuity depends largely on "frame of mind". It appeared that the more the subjects showed anxiety about how they were doing, the worse they did. The routine analysts were used to doing tests, exercising the greatest possible care, but generally without even being interested in the purpose of the work.

Subjects were given a pair of cylinders (diameter 2 cm, height 2.5 cm), one in each hand, and asked to squeeze them twice between finger and thumb,

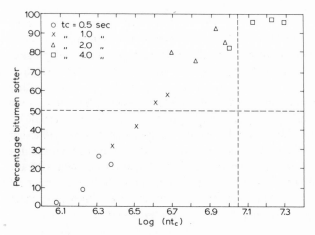

Fig. XI (2). Manual comparisons of softness of bitumen and rubber samples. Reproduced from: Scott Blair, G.W. and Coppen, F.M.V., Amer. J. Psychol., 55: 215 (1942).

change hands (to eliminate effects of left- and right-handedness), to attempt to exert a steady pressure, and then to report on which sample felt softer. Adequate temperature control for the bitumen samples was maintained by keeping the hands and the samples under water at constant temperature. "Softness" was preferred to "firmness" because of possible confusion between "firmness" and "hardness". It was assumed, at first, that the softness judgements were made by an estimate of *amount* of compression and the viscosity judgements by estimating *rate*.

To follow up such speculations, a third series of experiments was done in which a viscous sample was directly compared with an elastic cylinder. Squeezing time was restricted by means of a metronome to four periods: 1/2, 1, 2 and 4 seconds. Apart from one or two subjects trained in physics who were clearly worried by the apparent impossibility (on dimensional grounds) of describing an elastic modulus as greater or less than a viscosity, subjects showed no difficulty in reaching decisions when differences were above the threshold (see ref. 45, third paper).

In a later paper [46] it was shown that a plot of log (nt_c)* vs. percentage "bitumen softer" answers gave a curve which, although ultimately no doubt sigmoid in character, was remarkably linear and independent of the value of t_c over most of the range. The curve is shown in Fig. XI-2.

---

* t_c is the time of squeezing; n is the stiffness of the spring.

The question arose: "by what are subjects judging softness when they compare a viscous and an elastic material in time to a metronome?"

The author was working, at this period, on the rheology of soft plastics and was able to obtain a series of materials for which the stress–strain relation was linear, but the strain–time curves followed a power equation. Many further "squeezing" experiments, which need not be described here, were done comparing the subjectively assessed "softness" of these materials as against steel springs, and a rather strange conclusion was reached, all other possible explanations proposed being eliminated by the results of experiments.

In a book concerned with biorheology, it is perhaps best to start with the conclusion and to work backwards to the equations; though the mathematical treatment will not be given. The fundamental aspects were dealt with in three papers by the present author and his colleagues [45, 46, 47].

Subjective assessments of time interval equality

When the physicist says that two objects are of equal length, he means, in principle, that they can be laid down alongside one another so that the ends coincide. When he says that two masses are equal, a simple way of expressing his meaning is to say that the two weights "balance", under the influence of gravity, when placed on the beam of a balance, equidistant from the fulcrum. But to say that two times are of equal duration involves more complex definitions, because times cannot be laid down alongside one another.

A good definition would be that intervals of time are equal when light, not in the proximity of matter, takes these times to traverse equal (i.e. theoretically superposable) intervals of space; and there are also quite a number of physical phenomena that show a periodicity corresponding, either exactly or very closely, to this "Newtonian time scale"*.

But we do not have "Newtonian clocks" in our brains and much work has been done by psychologists and physiologists on trying to find out how we judge time intervals. It is evident that, day-to-day, time seems to pass in a most irregular manner. When we are enjoying ourselves, or asleep, time passes quickly: when we are anxious or bored, very slowly. Taken over a lifetime, it seems as if something more like a logarithmic time scale applied. As I look back to my earliest memories, the years seem to have been increasingly long: in old age, they pass with increasing speed.

---

* Milne's "Kinematic Relativity" postulates a very small difference between atomic and macroscopic time scales. The time scales are related logarithmically. Many authors have proposed a logarithmic scale for biological time.

With the invention of the differential calculus by Newton and Leibnitz, a system already certainly existent in embryonic form, was made quantitative. One has l (a length), dl/dt (a velocity), d^2l/dt^2 (an acceleration), etc. Indeed, differentiation with respect to time is so common that a simple notation used by Newton—"Newton's dot"—is still used, e.g. $\dot{\gamma}$ is a common abbreviation for $d\gamma/dt$. We have already mentioned the rather obvious suggestion that subjects squeezing springs at, as far as possible, a constant pressure, will judge "stiffness" by a strain (a relative length) and those judging viscous systems, by a rate of change of strain.

Scott Blair and Coppen [45] proposed that, when both judgements had to be made simultaneously, or when visco-elastic materials are being judged, the criterion will lie between a length and a velocity and could be expressed in terms of a fractional differential of the form $d^n l/dt^n$ ($1 > n > 0$).

Fractional differentials were already well-known to mathematicians, notably through the work of Heaviside. They can be expressed in the form of a modification of the binomial theorem to include fractional indices, known as "gamma functions". Although there are serious difficulties in defining fractional differentials, in such a way as to satisfy the mathematicians, the equations can be integrated to give a series of terms, the first of which is a simple power relation*. Scott Blair and Coppen [45] showed that, for materials linear with respect to stress–strain (γ) but giving good straight lines for log strain vs. log time, the slope of these lines agreed, within experimental error limits, with the exponent of the first term of the integrated equation derived from statistical treatment of data obtained by purely subjective comparisons of "softness" with that of springs or bitumen samples.

Harper [48], at that time working with the author, made considerable advances in this difficult field by applying "probit analysis" to the curves that Scott Blair and Coppen [45] had regarded as adequately linear on a logarithmic plot. This method treats them as sigmoid (as indeed they are) and Harper [48] was able to show that the fractional exponent of differentiation from the subjective data is indeed consistent with the slope of the log γ/ log t curves obtained using a rheometer.

The author has found it difficult to summarize this early work intelligibly without introducing the equations. It is hoped that readers will have gained some idea of the significance of these ten years of experimentation in showing how the brain quantifies changes in length with respect to time.

The alternatives would seem to be to use a non-Newtonian time scale (which the author has been wrongly accused of doing) or, keeping the ordi-

* The author was indebted to his friend the late Dr. Paul White for many of the calculations.

nary scale of time equalities, to introduce fractional differentials. Adequate definitions of the latter would not appear yet to have been introduced, at least in connection with this problem.

After the ending of this programme, little work on these lines appears to have been done in this field until quite recent times, although many industrial rheologists and psychologists have made direct comparisons between subjective assessments of rheological properties and data from suitable rheometers, without introducing any specific theory.

However, a more fundamental study is now being made in Japan. The work is still at an early stage, and only one published paper can be quoted at the time of writing (Morosawa et al. [49]). Future publications will be awaited with great interest.

REFERENCES

[1] Bergson, H., *Essai sur les Données Immédiates de la Conscience* (Paris, 1889) (English transl., *Time and Free Will* (H. Pogson) (Swan Sonnerschein and Co., London, 1910)].

[2] Stevens, S. S., *Science*, 103: 677 (1946).

[3] Stevens, S. S., in *Measurement, Definitions and Theories* (Eds. C. W. Churchman and P. Ratoosh), Chapter I (Chapman and Hall, London, 1961).

[4] Stevens, S. S., *Amer. Scientist*, 48: 226 (1960).

[5] Stevens, S. S., *Science*, 133: 80 (1961); also many other papers.

[6] Treisman, M., *Brit. J. phil. Sci.*, 13: 130 (1962).

[7] Treisman, M., *J. acoust. Soc. Amer.*, 42: 586 (1967); also many other papers.

[8] Scott Blair, G. W., *J. Soc. cosmetic Chem.*, 17: 45 (1966).

[9] Harper, R., *Lab. Pract.*, 7: 578 (1958); 7: 648 (1958); 7: 712 (1958).

[10] Scott Blair, G. W., *A Survey of General and Applied Rheology*, Chapter XVI (Pitman, London, 2nd edn., 1949).

[11] Sullivan, A. H., *Amer. J. Psychol.*, 33: 121 (1922) and 34: 531 (1923).

[12] Sullivan, A. H., *J. exp. Psychol.*, 10: 147 (1927).

[13] Zigler, M. J., *Amer. J. Psychol.*, 34: 73 (1923).

[14] Meenes, M. and Zigler, M. J., *Amer. J. Psychol.*, 34: 542 (1923).

[15] Binns, H., *J. Text. Inst. Manchr.*, 29: T 117 (1938).

[16] Harper, R., McKennell, A. C. and Onions, W. J., *J. Text. Inst.*, 49: 126 (1958).

[17] Katz, D. and Stephenson, W., *Brit. J. Psychol.*, 28: 190 (1937).

[18] Katz, D., *Der Aufbau der Tastwelt* (Barth, Leipzig, 1925).

[19] Renner, H. D., *Origins of Food Habits,* Chapter IV (Faber and Faber, London, 1944).

[20] Stevens, S. S. and Guirao, M., *Science*, 144: 1157 (1964).

[21] Fryklöf, L. E., *Särt. Svensk. farmaceut. Tidskr.*, 63: 697 (1959).

[22] Yoshida, M., *Jap. psychol. Res.*, 10: 123 (1968) and 10: 157 (1968).

[23] Katz, D., *Cereal Chem.*, 14: 382 (1937).

[24] Katz, D., *Occup. Psychol.*, 12: 139 (1938).

[25] Scott Blair, G. W. and Potel, P., *Cereal Chem.*, 14: 257 (1937).

[26] Marriott, R. H., *The Analyst*, 74: 397 (1949).

[27] Prentice, J. H., *Lab. Pract.*, 3: 186 (1954).
[28] Uzzan, A. and Sambuc, E., *Rév. franç. Corps Gras*, (1959).
[29] Naudet, M. and Sambuc, E., *Rév. franç. Corps Gras*, (1959).
[30] Matz, S. A., *Food Texture* (Avi Publ. Co., Westport, Conn., 1962).
[31] Sone, T., *Consistency of Foodstuffs* (D. Reidel Publ. Co., Dordrecht, The Netherlands, 1972).
[32] Parkinson, C. and Sherman, P., *J. Texture Stud.*, 2: 451 (1971).
[33] Moskowitz, H. R., Drake, B. and Åkesson, A. J., *J. Texture Stud.*, 3: 135 (1972).
[34] Scott Blair, G. W., Chapter VIII and Harper, R., Chapter IX, in *Some Recent Developments in Rheology* (Ed. V. G. W. Harrison) (United Trade Press, London, 1950).
[35] Eysenck, H. J., *The Dimensions of Personality* (Kegan Paul, London, 1947).
[36] Harper, R., *Amer. J. Psychol.*, 60: 554 (1947).
[37] Harper, R., *New Scientist*, 11: 396 (1961).
[38] Szczesniak, A. S., *Food Technol.*, 22: 981 (1968).
[39] Stone, P. T., *Automotive Body Engng*, 135: 28 (1965).
[40] Drake, B., *J. Food Sci.*, 28: 233 (1963).
[41] Drake, B., *Biorheology*, 3: 21 (1965).
[42] Drake, B., *Proc. 3rd int. Congr. Food Sci.*, Washington, D. C., p. 437 (1970).
[43] Scott Blair, G. W. and Coppen, F. M. V., *Nature*, 143: 164 (1939), 144: 286 (1939) and 145: 425 (1940).
[44] Scott Blair, G. W. and Coppen, F. M. V., *Brit. J. Psychol.* (*gen.*), 31: 61 (1940).
[45] Scott Blair, G. W. and Coppen, F. M. V., *Amer. J. Psychol.*, 55: 215 (1942).
[46] Scott Blair, G. W., Veinoglou, B. C. and Caffyn, J. E., *Proc. roy. Soc. A*, 189: 69 (1947).
[47] Scott Blair, J. W. and Caffyn, J. E., *Phil. Mag.*, 40: 80 (1949).
[48] Harper, R., *Amer. J. Physiol.*, 60: 554 (1947).
[49] Morosawa, K., Ohtake, C., Takahashi, M., Mitsui, T. and Ishikawa, S., *J. Soc. cosmetic Chem.*, (1973).

BOOK FOR FURTHER READING

Harper, R., *The Human Senses in Action* (Churchill Livingstone, Edinburgh–London, 1972).

Chapter XII

BOTANICAL ASPECTS OF RHEOLOGY

by
D. C. SPANNER

Compared with the animal world, the world of plants has relatively little to offer in the way of rheological interest; that is, if we exclude the phenomena associated with protoplasm itself (see Chapter III). This is not surprising in view of the much more sedentary life that plants live. Nevertheless the movement of liquid material in plants is an important and large scale one, and as such has inevitable interest from the rheological point of view. The liquids which move, with the notable exception of latex, are, however, usually much simpler than those (such as blood) which interest the animal physiologist. This orients the rheological interest rather differently, but the phenomena of flow still raise very interesting problems. In the present introduction the botanical aspects will be dealt with under three headings: movement in the xylem, movement in the phloem, and latex flow. A final brief section will deal with some miscellaneous topics.

Flow in the xylem

The upward movement of the watery sap in the woody tissues (xylem) of trees and lianes is a phenomenon of very considerable magnitude. While its chief function seems to be making good the water lost by evaporation from the leaves it nevertheless fulfils other functions as well, notably the transport upwards from the roots of mineral ions and organic nitrogen. This transport may represent their initial delivery to the aerial portions of the plant, or it may form part of a recirculation of nutrients within the plant, a process which is now recognized as being quite widespread and significant (Biddulph [1], Crafts and Crisp [2]). However, from the rheological point of view the chief interest lies in the dynamics of the flow process, and it is on this aspect that attention will be concentrated.

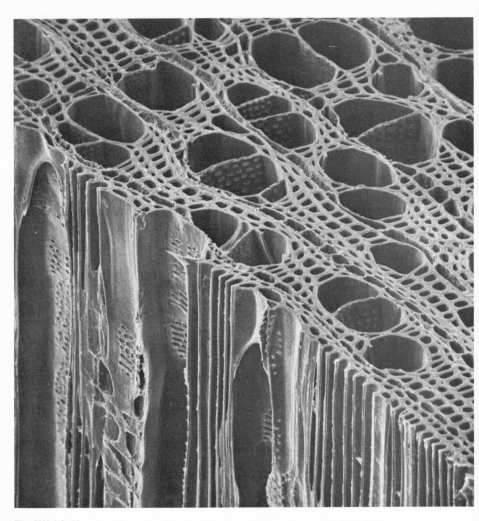

Fig. XII (1). Vessels and smaller-diameter fibre-tracheids in the wood of *Nothofagus fusca*. Oblique rims representing the remains of the end-walls of vessel elements can be seen on the vertical face. x 300. Reproduced by courtesy from: B.A. Meylan and B.G. Butterfield. Three-dimensional structure of wood, Chapman Hall, London (1972).

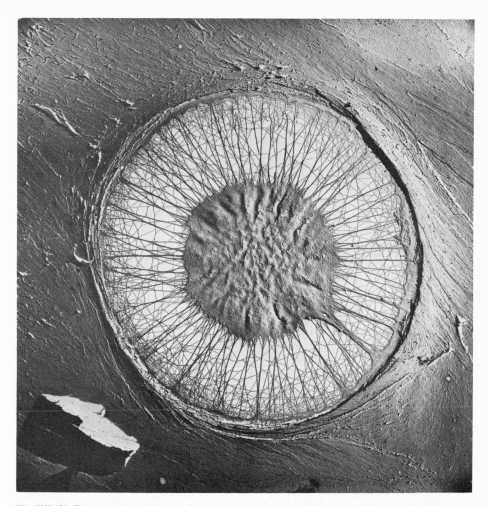

Fig. XII (2). The complex membrane that separates the opposing pits in contiguous tracheids of *Abies grandis* (Fir). Water passage is through the open-textured annular region of the membrane. x 8100. By courtesy of G.S. Puritch and R.P.C. Johnson.

Conducting elements

The conducting elements of the wood of trees (Esau [3]) are of two broad types. In the more advanced classes—the flowering trees represented by the oak or the palm—the main channels are made up of relatively short dead, lignified cells arranged end-to-end and with the cross walls broken down to form large open pores (Fig. XII-1). What remains of the cross wall may be pierced with one very large, or numerous smaller perforations; but in every case there is an obvious continuous channel formed for fluid movement. Such open channels (of considerable length) are spoken of as vessels or tracheae, and their diameter may vary typically from say 50μm in "diffuse-porous" trees to 400μm in "ring-porous" trees (where the large vessels are visible to the naked eye as a "ring" in the spring wood). In the less advanced classes—the conifers represented by the pine and cedar, for instance—the conducting channels are composed of dead elongated cells known as tracheids, with tapered ends. These communicate with one another not through relatively large pores opened through the walls, but through very numerous small thin specialized areas called "pits" where the wall is reduced to a rather permeable structure of cellulose microfibrils. The pits are, in fact, rather complex in design, and include a built-in mechanism which seals them in the event of too large a pressure difference developing between adjacent cells (Fig. XII-2). Tracheids have typically a diameter of 40 to 60μm and a length of 3 to 5 mm. The vessels, on the contrary, have lengths of anything up to 20 m as judged by their ability to permit passage of air through the cut stems under low pressure.

Velocity of sap stream

It can thus be seen that the water-conducting tissues in trees possess a structure which, a priori, would seem to offer a wide range of resistances to the flow of the sap. In keeping with this expectation measurements of the stream velocity vary from about 1 metre per hour in diffuse-porous trees with small vessels, to about 45 metres per hour in ring-porous trees with large vessels, though some of these measurements made by observations on the movement of a heat-pulse in the stem suffer from certain theoretical difficulties that make them underestimates (see Marshall [4]). However, the interesting, but not unexpected point emerges that the velocity depends very markedly on the diameter of the vessels. This of course is in accordance with the Poiseuille equation for a circular pipe, which may be put in the form:

$$v = \frac{r^2}{8\eta} \cdot \frac{\triangle P}{\triangle x}$$

where v is the mean linear velocity, r the radius, η the viscosity and $\triangle P/\triangle x$ the pressure gradient; the only proviso in this conclusion is that the pressure gradient does not vary markedly from one species to another. Since the total pressure differential available depends ultimately on the osmotic pressure of the cell sap in the mesophyll of the leaves this is not an unreasonable assumption for trees of a similar stature.

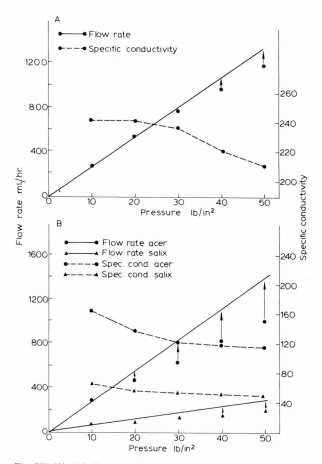

Fig. XII (3). (A) Relationship between flow-rate, specific conductivity, and the pressure difference between the ends of the xylem cylinder (*Fraxinus*). (B) Relationship between flow-rate, specific conductivity, and the pressure difference between the ends of the xylem cylinder (*Acer* and *Salix*). Reproduced by courtesy from: Peel, A.J., Annals Bot., 29:119–130 (1965).

Adequacy of the Poiseuille equation for the xylem

From the rheological point of view an interesting question relates to the adequacy of the Poiseuille equation to describe water movement in the xylem. Peel [5] tested this on three species of tree, one (*Fraxinus*) having ring-porous wood (i.e. large vessels) and two (*Salix*, *Acer*) being diffuse-porous. Pieces of stem 30 cm long were used and water or glycerol solutions forced through under pressure. The results were interesting. With *Fraxinus* the volume flow rate was proportional to the pressure difference up to a value of about 30 lb. in^{-2} (2 atmos); after that it fell below proportionality (Fig. XII-3A). With *Salix* and *Acer*, which have much smaller vessels, the fall below proportionality occurred much earlier (Fig. XII-3B). Peel interpreted this departure from Poiseuille behaviour as due to turbulence. The large vessels of ring-porous trees are known to be much longer than the narrow ones of diffuse-porous species. If the end walls introduce turbulent flow this would therefore be more conspicuous with the latter type and the effect would be of the kind experienced. Peel did not carry the inquiry further by looking into the Reynolds numbers; clearly this would be desirable in future work.

When solutions of raised viscosity were forced through the stems the results diverged more widely from predictions. According to the equation a plot of volume flow rate against the reciprocal of the viscosity should be linear; in fact it was markedly curved with both types of stem (Fig. XII-4). In all of the curves presented here the specific conductivity (defined as the volume flow rate × viscosity/pressure gradient × cross-sectional area of stem) shows the departure from Poiseuille behaviour very clearly; ideally it should remain constant for a given specimen. It seems possible that the unexpected effect of a rise in viscosity is due to an osmotic interaction of the glycerol solution with the living cells in the xylem, or to an effect on the imbibition of the colloidal material of the vessel walls, though it is not easy to suggest a precise mechanism. The xylem, however, is a complex tissue, and the results are a reminder of this fact. Further experiments in which flow was measured under low pressure (10 lbs . in^{-2}), the stem then subjected to high pressure (50 lbs . in^{-2}) and the flow re-measured under low pressure showed that the high-pressure treatment, repeated several times, progressively lowered the conductivity to a steady value. The explanation advanced by Peel was that under high pressure the water dissolved more air. On return to low pressure bubbles were formed and the smaller channels became blocked by air-embolism. Whether this be the correct explanation or not the complexity of the xylem as a flow channel is again emphasized.

One further point re-emerging from the same author's work concerns the

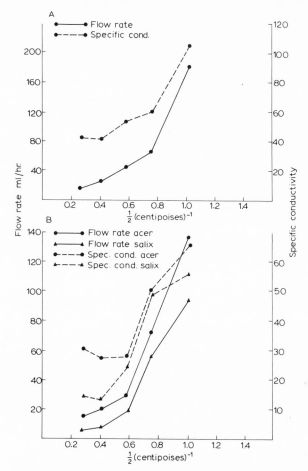

Fig. XII (4). (A) Relationship between flow-rate, specific conductivity, and the reciprocal of the viscosity of the xylem liquid (*Fraxinus*). Pressure 10 lb/in.² (B) Relationship between flow-rate, specific conductivity, and the reciprocal of the viscosity of the xylem liquid (*Acer* and *Salix*). Pressure 10 lb/in.² Reproduced by courtesy from: Peel, A.J., An. Bot., 29: 119–130 (1965).

radial extent of the functional xylem. This question is not, perhaps, strictly rheological, but it is worth noting here; progressively, the older xylem becomes blocked by balloon-like ingrowths into the vessel cavities from the neighbouring living cells. These tyloses, as they are called, put the vessels out of use, and as a consequence the radial extent of the conducting xylem may be fairly small. In ring-porous trees it may be only the outermost annual ring which is conducting; in diffuse-porous trees considerably more are involved.

Hydraulic conductivity of wood

Recently the idea of the water conductivity of the xylem has been put on a rather firmer basis. Heine [6] has collected numerous published figures relating to flow velocity and water conductivity. He analyses these and in particular calculates from them the pressure gradient operative at the velocities quoted. For *Quercus* (ring-porous) the gradient may be about 0.8 bar.m^{-1}; for *Betula* (diffuse-porous) about 0.63 bar.m^{-1}; and for 'small conifers' (with tracheids) about 0.6 bar.m^{-1}. The correspondence of these values in widely different types of woody tissue is indicative of the importance of the (radius)^4 term in the Poiseuille equation.

Motive force for the ascent of sap

It has long been held as more-or-less established that the xylem sap is drawn upwards in the vessels by a tension from above, sustained by the cohesive strength of the sap and its adhesion to the vessel walls. Details of this so-called "cohesion theory" can be found in standard textbooks of plant physiology; an excellent review is given by Zimmermann [7]. From the present point of view interest lies in the implication that the water columns are usually under considerable tension. An interesting example of this is the finding by Scholander and his colleagues [8] that the tension in the xylem columns of certain mangroves is regularly between 30 and 60 atmospheres, this high negative pressure being associated with the extraction of fresh water from the sea water which bathes their roots by a process of reverse osmosis. A high negative pressure, however, makes the water metastable, and any suitable nucleation will result in the formation of vapour bubbles leading to air embolism in the channels. This is evidently a serious problem for tall trees, for once the water column is broken the vessel is useless. Clearly the larger (and so probably longer) the vessel the more serious is the danger; and it seems likely that this fact offsets the desirability, from the point of view of low resistance, of large vessels. Ring-porous trees do not, on this account, seem to rely on last year's vessels; they develop new ones in advance of the emergence of leaves. Thus, as Zimmermann points out, in these vulnerable trees air embolism caused by freezing of the xylem contents in winter (which throws air out of solution) does not leave the newly-emerging spring leaves without a water supply. Certain diffuse-porous trees, such as the birch and maple, adopt a different plan. Their vessels are narrower, shorter and more numerous. Early in spring, and before the new wood is developed, root pressure thrusts the sap up from beneath and many air pockets are dissolved, rehabilitating the vessels. Finally, the pines and other conifers of the cold regions manage to survive freezing of their trunks because their wood contains not vessels, but only tracheids. The bubbles of air thrown out of solution remain dispersed in individual elements, unable to coalesce. Slowly their oxygen is used in respiration; on thawing in the spring, therefore, the small bubbles are the more readily able to redissolve before the transpirational demand of the leaves begins.

The rheological situation in the xylem

Broadly, the situation in the xylem of trees can be characterized in this way. It seems to reflect the Poiseuille relationship, and as such the two factors of importance are the diameter of the conduits, and the maintenance of a pressure gradient; the viscosity of the sap has not been widely investigated and would seem to be of minor interest. Clearly, the wider the conduits the better from the point of view of function. On the other hand, the wider the conduits the more is the continuity of the water column at risk, and with it the maintenance of the driving force. A compromise is the inevitable result, humanly speaking; and the nature of the compromise depends on the climatic and soil conditions in which the tree has to flourish.

Movement in the phloem

The other considerable movement in plants occurs in the phloem, a tissue which almost invariably runs parallel and close to the xylem on the outside. It distributes not water but nutrients, mainly the products of photosynthesis elaborated in the leaves. The assimilate stream, as it is called, contains sugar (most often sucrose) as by far its most abundant solute, though mineral ions and many other constituents are important. Frequently the current in the phloem runs in the opposite sense to that in the xylem (though it is capable of altering direction); as a consequence many solutes in the plant are subject to recirculation [1, 2]. Since carbohydrate material in the form of cellulose, forms the bulk of the substance of the plant body, and all of it originates in the leaves, it can be seen that the phloem movement is a very substantial one in terms of the amount of substance it transports, though of course purely as a movement it is not so considerable as that in the xylem.

174

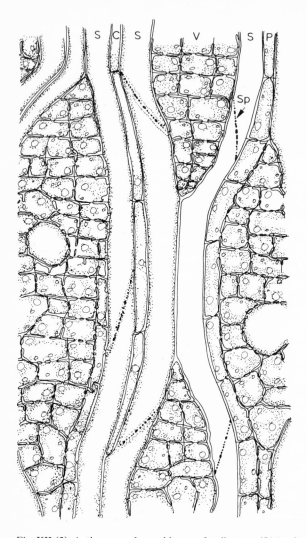

Fig. XII (5). Active secondary phloem of tulip tree (*Liriondendron*) in tangential section, showing sieve tubes (S), with sieve plates (Sp), at the ends of sieve tube elements; C, companion cells; V, vascular ray (phloem ray); P, phloem parenchyma. The walls of cells in the vascular ray and phloem parenchyma show numerous simple pits. Sieve tubes are about 40 μ in diameter, pores about 2 to 5 μ. Reproduced by courtesy from: O.F. Curtis and D.G. Clark, An Introduction to Plant Physiology, McGraw Hill, New York (1950).

The phloem conduits

Interesting as are the xylem channels from the rheological angle, the phloem conduits are much more so. In general shape they are roughly similar, but smaller, having diameters of from 10 to 50 μm, i.e. about one quarter those of vessels. A major point of difference is that the phloem elements are living cells, not dead ones, though their protoplasmic contents are reduced to a very thin peripheral layer. The elements, which may be from 100 to 500 μm long, are arranged in long files like the vessel elements; and in the flowering plants the end walls between the individual cells (again recalling the xylem vessels) are perforated with open pores. These are not nearly so large as the corresponding openings in the vessels, ranging usually from 1 to 4 μm in diameter; but they are very conspicuous, and give to the end walls their well-known name of "sieve plates", and to the phloem conduits, "sieve tubes" (Fig. XII-5). The pores in the sieve plates evidently constitute part of the overall flow path, and it is important to note that they are not crossed, but rather lined, by the vital and highly impermeable plasma membrane. Were it otherwise, i.e. did the membrane seal off the sieve plate pores, the resistance of the plates to movement of the assimilate stream would be many orders of magnitude higher.

Besides the sparse cytoplasm, a few mitochondria, and rather more numerous peripheral starch-containing plastids, the sieve tubes contain as their most conspicuous component an insoluble fibrillar protein formerly referred to as "slime" but now named P-protein. This looks remotely like collagen in some of the latter's forms (see Chapter IV). It consists of long filaments with a repeating structure composed apparently of two threads twined helically. In most electron micrographs of sieve tubes it appears in the vicinity of the sieve plates, almost always filling the sieve plate pores and aggregated on both sides (Fig. XII-6). Often it is drawn out in long strands in the lumen running to the pores. The diameter of the individual filaments is about 100 Å, and where the filaments run together through the sieve plate pores their distance from one another seems to be a matter of a few times this diameter only.

In the gymnosperms (e.g. the pines) where tracheids replace vessels, a comparable distinction appears in the phloem elements. The pores are very much smaller than in the flowering plants and technically files of "sieve cells" replace "sieve tubes".

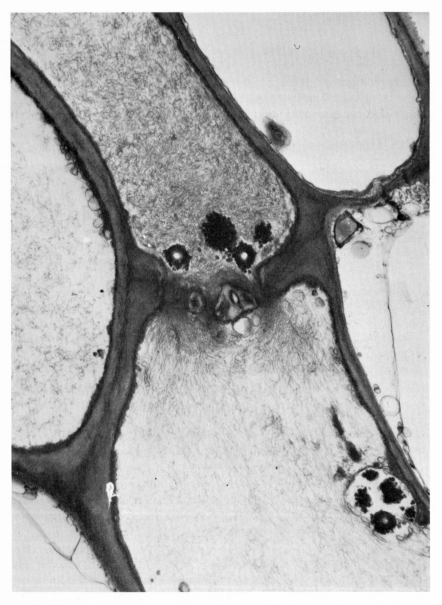

Fig. XII (6). Sieve tubes and sieve plate from the stolon of *Saxifraga sarmentosa*. The fibrous material in the cells is P-protein; the dark globular bodies are starch grains. ×20,000. By courtesy of A.W. Siddiqui.

Hydrostatic state of the conduits

Before further discussing the rheology of the problem an important distinction between the xylem and the phloem must be made. It has been pointed out that the xylem vessels commonly sustain a considerable negative pressure, the transpirational tension. The phloem elements, on the contrary and owing to the high solute concentration in their sap, sustain a high positive pressure, perhaps 20–30 atmospheres. It is this circumstance that has led to much uncertainty about their native state.

State of the functioning sieve element

The resistance to movement in the sieve tubes is made up of the flow-resistance in the lumens of the sieve elements together with the resistance to passage across the sieve plates. These two components would be comparable were it not for indications that the sieve plate pores are occluded with filaments of the P-protein. Long ago light-microscope images of the sieve tubes invariably showed a cone-shaped mass of amorphous material (the "slime plug") clinging to and penetrating the sieve plates. Electron micrographs have refined this picture but, many would maintain, not substantially altered it. The sieve plate pores appear, generally, to be occupied by filaments neatly packed into and running through them in a configuration recalling stream-line flow. Sometimes they communicate with masses of dense P-protein lying against either side of the plate (Fig. XII-6); sometimes the material in the pores stands alone in its density (Fig. XII-7). The highly controversial question is thus raised: Is this apparent occlusion of the pores with filamentous protein an artifact? Those who argue that it is do so on the basis that any interference with the sieve tubes (like excision) immediately results in violent surges on account of the high pressure and longitudinal continuity. These surges carry P-protein dispersed in the lumens into the sieve plate pores and compact it there, producing an image quite different from the natural reality. In vain has every conceivable method been tried to avoid these surges in an incontrovertible fashion; the controversy remains unsettled. For what it is worth, the present author's opinion is that the indications of micrographs such as Figs. XII-6 and 7 are to be accepted at their face value, and that any theory of the translocation process must come to terms with occluded pores. The use of freeze-etching technique supports this opinion; but as will be seen in a moment the rheological problems involved if this position be accepted are very great indeed.

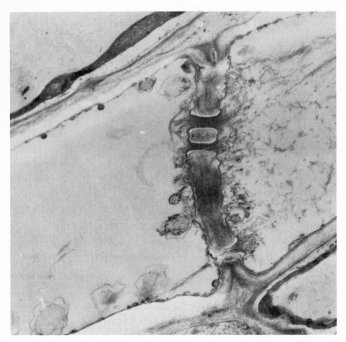

Fig. XII (7). Sieve plate from petiole of tomato. P-protein occupies the pores. The thin white lining of the pores is callose. ×17,500. Reproduced by courtesy from Yapa, P.A.J. and Spanner, D.C., Planta, 107: 89–96 (1972).

The problem of phloem movement

To indicate the problem certain additional facts need to be remembered. The assimilate stream consists of a solution containing from 10–25% sugar together with about one-tenth as much amino acids, and rather less potassium and other small ions. Its viscosity might therefore be about 1–5 centipoises. Further, the velocity of movement has been established by a variety of methods as of the order of 30–200 cm·h^{-1}; this may be taken as the mean velocity in the lumen, and it would be perhaps double or treble this in the pores of the sieve plates. On the basis of Poiseuille's equation therefore it can easily be calculated that a pressure gradient of the order of 0.5–1 atmosphere per metre would be required to propel the stream through sieve tubes entirely empty (Weatherley and Johnson [9]). But the matter is very different indeed if the sieve plate pores are occluded. Calculation then shows that the pressure gradient required may easily amount to several hundred atmospheres per metre, a magnitude obviously quite impossible with a total turgor pressure of only 20–30 atmospheres. Since this conclusion is so important to the whole problem its basis is worth setting out more clearly.

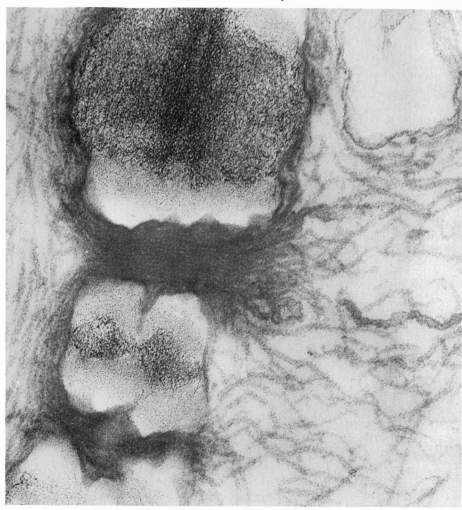

Fig. XII (8). Sieve plate pores from the petiole of *Nymphoides*. The P-protein fibrils are closely-packed in the pore, which is lined by the plasmalemma membrane. The electron-lucent areas adjacent to the pore are callose. x 102,000. By courtesy of R.L. Jones.

180

Flow along an array of filaments

The most resistive part of the channel is of course (on the present understanding of Figs. XII-6 and 7) the pores; and in the pores the fibrils seem to
run in a close, almost semi-crystalline array (Fig. XII-8). The situation can
be approximated for mathematical purposes by a regular array of hexagonally-
arranged filaments parallel to which flow occurs (Fig. XII-9). The channel is
divided into hexagonal flow elements each centred on a filament of radius ρ.
If these are then imagined to be replaced with cylindrical elements (radius R)
of equal cross-section the problem is greatly simplified, the boundary
conditions for a Poiseuille-type analysis being simply $v = 0$ for $r = \rho$, and
$dv/dr = 0$ for $r = R$. This leads very simply (Fenson and Spanner [10]) to
an expression for the axial pressure gradient:

$$\frac{\triangle P}{\triangle x} = \frac{8\eta v}{R^2} \bigg/ \left\{ \frac{4\ln/\alpha}{1-\alpha^2} - (3-\alpha^2) \right\}$$

where v is the mean velocity, $\alpha = \rho/R$, and the spacing of the filaments
centre-to-centre is 1.90 R. Taking as typical values $2\rho = 100$ Å, fibrillar
spacing 1.90 $R = 600$ Å and sieve plate thickness $\Delta x = 1$ μm, with a pore
velocity of 100 cm·h⁻¹ the equation gives $\triangle P = 0.007$ atmosphere for the
pressure drop over a single sieve plate. With 2500 plates per metre this involves
a pressure drop of 17 atmospheres per metre, impossibly high if the flow is
to be pressure-driven in a tree of any size. It is this sort of consideration that
argues very strongly for the artifact-nature of the image presented by such
micrographs as Figs. XII-6, 7. The values of v, η, ρ and R which have been
taken are by no means pessimistic ones; most micrographs in fact suggest
that R may be considerably smaller, and indeed Fig. XII-8 does so.

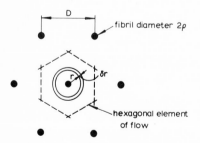

Fig. XII (9). Cross section of ideal hexagonal array of parallel fibrils running through a sieve
plate pore. Reproduced by courtesy from: Fenson, D.S. and Spanner, D.C., Planta, 33:
321–331 (1969).

Validity of macroscopic equations

Implicit in the above calculations is the assumption of the validity of the macroscopic approach to a system whose scale is very small indeed. A hydrated sucrose molecule may be 9 Å in diameter; the P-protein fibrils are only 100 Å, and the spaces between them not very much greater. In the circumstances does the use of an equation based on the macroscopic viscosity lead to wildly inaccurate results? At least this question is worth a comment.

There are two fairly obvious approaches possible. The first is illustrated by the work of Churayev, Sobolev and Zorin [11]. Experimenting with extremely fine bore quartz tubes these workers determined the viscosity of various liquids using Poiseuille's equation. They found that non-polar liquids like carbon tetrachloride showed a quite normal viscosity in capillaries down to 0.08 μm in diameter; with water, however, the viscosity in this size of capillary appeared to be 40% above the usual value. However, an alternative interpretation is possible, they state. The viscosity of the water may have been quite normal, the apparent rise being due to the absorption on the capillary walls of a fixed aqueous film of about 80 Å thickness. For the present purpose it hardly matters which is the right way of viewing things. The important point is that down to a diameter of 0.08 μm or 800 Å the use of the Poiseuille equation is not wildly inaccurate; in fact it seems to underestimate rather than overestimate the resistance.

The second approach to the problem is by way of the Stokes' equation (see Chapter II). This gives the resistance to movement of a solid sphere in a viscous fluid. Now it is possible to find ions singly-charged, symmetrical and large enough to be virtually unhydrated. The electrical mobility of these can be measured and compared with the value calculated from the Stokes' equation using radii obtained from well-known atomic dimensions. Such ions

TABLE 1

Ion	r (Å)	r_S (Å)	r/r_S
$N(CH_2)_4^+$	3.47	2.05	1.69
$N(C_2H_5)_4^+$	4.00	2.82	1.42
$N(C_3H_7)_4^+$	4.52	3.93	1.15
$N(C_4H_9)_4^+$	4.94	4.73	1.04
$N(C_5H_{11})_4^+$	5.29	5.27	1.00

r = radius estimated from molecular volumes or models
r_S = radius calculated from the limiting mobility by Stokes' law.

182

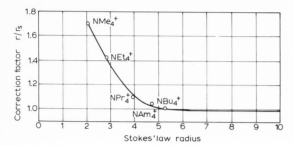

Fig. XII (10). Tentative correction factors for Stokes' law in water at 25°. Reproduced by courtesy from: Robinson, R.A. and Stokes, R.H., Electrolyte solutions, Butterworths, London (1970).

are the substituted quaternary ammonium ions, and the data in Table XII-1 and Fig. XII-10 taken from Robinson and Stokes [12] seem to indicate that at least down to the size of the quaternary amyl ion the use of the macroscopic Stokes' equation gives quite accurate results. This carries the matter considerably further down towards the molecular level than the "capillary" approach, and both lines of evidence seem to agree in the conclusion that there is no escape from the problem of occluded pores by supposing that the resistance they offer to sap movement is much less than that suggested by the use of the Poiseuille and similar equations, unless indeed, as despairing plant physiologists have sometimes suggested, molecular "slip" or some quite special lubricating properties at the surface of the P-protein fibrils can be invoked to reduce the flow-resistance drastically.

The problem of mechanism

The problem therefore is to conceive of a mechanism capable of propelling the assimilate stream through occluded pores at the relatively high velocity known from numerous experiments. This problem is hardly of direct rheological interest, and the briefest discussion must suffice. Besides those physiologists who cling to the idea that the functioning pores are quite open and that a simple pressure gradient is adequate to move the stream, there are others who accept the occlusion of the pores as genuine and for whom the answer lies in some sort of pumping action. The P-protein, they usually maintain, performs an essential role in this. One suggestion is that it has a mechano-chemical function like muscle (see Chapter IV) or other contractile protein; another is that it has a passive structural function as a charged framework for the development of electro-osmotic forces. An excellent review of the situation has been given recently by MacRobbie [13]; there is a shorter one by the present author (Spanner [14]). The helical structure of the P-protein (Fig. XII-11) might suggest either of these alternatives.

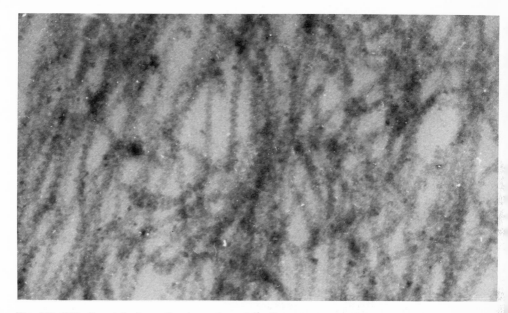

Fig. XII (11). P-protein from the sieve-tubes of *Saxifraga sarmentosa*. Note the helical structure, in places apparently double (see arrow). ×160,000. By courtesy of R.L. Jones.

Further comments

It seems fairly clear that the P-protein does not itself move with the assimilate stream. When functioning sieve tubes are severed by a razor cut drops of sap emerge from both sides, presumably under the thrust of the turgor pressure. The exuded sap contains little of the P-protein and the exudation commonly comes to a halt fairly quickly; it can be restarted by a second cut a short distance back from the first. Electron microscopic examination shows that the sieve plates near the cut bear obvious signs of turgor release artifacts; the P-protein has been forced through the pores and starch grains from ruptured plastids are usually very evident (Fig. XII-12). In addition a carbohydrate called callose (a β–1, 3 linked glucose polymer) is rapidly laid down on the sieve plate constricting the pores and helping to seal off the damaged

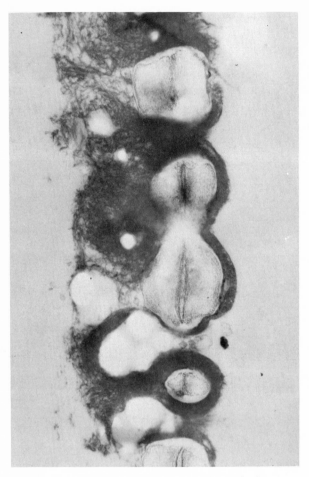

Fig. XII (12). Artificial displacement of P-protein by turgor release in sieve plate of *Heracleum*. The white bodies are starch grains. ×18,000. Reproduced by courtesy from: Mishra, U. and Spanner, D.C., Planta, 90: 43–56 (1970);

sieve tubes (Figs. XII-7, 8). These facts emphasize the vital nature of phloem function and the necessity of preventing loss of the valuable assimilate stream on injury. Callose formation is very rapid, the blocking of the pores by starch grains and P-protein even more so; but it is very doubtful if such a function of first aid to damaged sieve tubes is an adequate explanation of the ubiquitous presence of P-protein. Of interest in this connection is the fact that in some plants with very large sieve plate pores (e.g. *Cucurbita*) there is a soluble protein in the sieve tube sap which gels on exposure to air, apparently by the formation of disulphide bonds. This fast reaction obviously helps to prevent loss of the assimilate stream. Milburn [15] has reported the curious observation that firm finger massage of the stem of potted *Ricinus* plants for a few minutes a day for several days results in a considerably increased flow when the sieve tubes are subsequently cut. The explanation for this phenomenon is not clear, but it recalls the physical buffeting given to palms in the tropics to induce phloem exudation for the sake of their sugar. It has long been known, incidentally, that the stalks of the giant inflorescences of palms and such plants as *Yucca* exude phloem sap copiously if the young inflorescence is cut off, a fact made use of in commerce.

Latex

In certain families of flowering plants an opaque white or coloured juice exudes from cut surfaces, quite distinct from either the xylem or the phloem sap. Well-known examples are the white *latex* (as it is called, from the Latin for juice) of the poppy, and of the rubber tree. The former is commercially important because of the alkaloids it contains, the latter for its caoutchouc. The latex comes from a system of tubes which ramifies through the bark in trees or the cortex in herbs, and which arises in two different ways. In plants with latex *vessels* the walls between certain young cells break down much as in the development of the vessels of the wood except that an open much-branched and anastomosing network results in which, moreover, the cells remain living. In plants with latex *cells* the development is different; the young cells branch and grow intrusively at their tips so that each at maturity ramifies through the whole plant but remains a separate structure, i.e. no anastomosing occurs. In both cases, however, the result is similar: a system of tubes lined with a thin layer of cytoplasm and with the huge central vacuole filled with the characteristic latex.

The function of the latex system is still a matter for debate; excretion of waste products, storage, transport and defence against insects and animals have all been suggested. Rheologically it raises a number of interesting questions which are made more so by the obvious commercial factors involved; the exploitation of rubber is a case in point. The latex itself commonly contains salts, sugars, starch, alkaloids, tannins, gums, and caoutchouc among other things; and the more abundant smaller molecules in the presence of the protoplasmic membranes naturally give rise osmotically to a substantial turgor pressure, just as in the case of the sieve tubes. It is this turgor pressure which is responsible for the spontaneous exudation on injury.

Rheological properties of latex

The viscosity and flow properties of latex differ widely depending on the plant from which it comes. The latex of *Hevea brasiliensis*, the rubber tree of commerce, is a very sticky emulsion, and the viscosity rises very rapidly with concentration. This is important in practice since opening of the latex tubes at once causes a release of their turgor and a consequent fall in the water potential of their contents. Considerable absorption of water follows owing to the proximity of the xylem. This lowers the viscosity and so helps to promote latex exudation, an important effect. In contrast to the viscosity the surface tension of the latex (which controls drop size) changes little with concentration when this exceeds a certain minimum.

Recent work on the rheology of *Hevea* latex has revealed some interesting

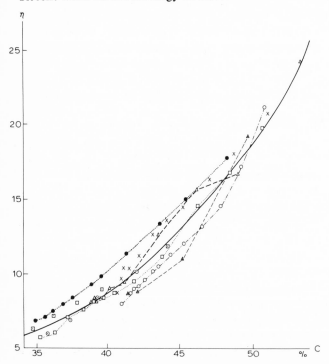

Fig. XII (13). Relative viscosity of *Hevea* latex. Abscissa, concentration c in % by weight of dry matter; ordinate, relative viscosity η. Reproduced by courtesy from: A. Frey-Wyssling, Deformation and Flow in Biological Systems, North Holland, Amsterdam (1952).

properties. Among the cytoplasmic contents of the latex tubes are some prominent globular bodies called lutoids. These are up to several microns in diameter, and are bounded by a membrane of molecular thickness. There is evidence that they contain a serum which can act as a destabilizer of the latex system, causing coagulation on rupture of the lutoids. Such rupture, which may be caused on tapping the trees by osmotic shock, shearing action or other means, might be an important reason for the gradual cessation of flow after the tapping operation. In fresh latex, lutoids have been shown to produce thixotropy when tested in a rolling-ball viscometer (see Chapter II). Fresh latex contains 30–40% by volume of the rubber hydrocarbon and about 10% of other particulate phases, notably lutoids. At one time the marked influence of the latter on the rheological properties was taken to indicate that they had a jelly-like interior: however, later studies have suggested that it is much more fluid. The well-known influence of additions of water or ammonia on the viscosity of fresh latex may be connected with the lutoids, water causing swelling and bursting (accompanied by a rise and then a fall in viscosity) and ammonia causing disintegration and a rise in pH (stabilizing the viscosity at a low level).

Further evidence for the influence of lutoids on viscosity has come from studies of capillary flow at high pressure gradients. When fresh latex was forced through small glass capillaries (22 to 80 μm diameter) gradients of 0.4 atm·mm^{-1} were usually about the maximum it could stand before flocculation brought flow to a standstill. When the lutoids were removed by centrifugation, however, much larger pressure gradients (up to 12.9 atm·mm^{-1}) could be tolerated. In capillaries of this size the apparent viscosity of fresh latex becomes less as the diameter is reduced, though removal of the lutoids lessens this effect. Since capillaries of the size under discussion are roughly the same diameter as latex vessels it is clear that it is hardly safe to apply the simple Poiseuille equation to them. Southorn and Yip claim that latex, flowing in a capillary, does not obey the r^4 law of Poiseuille. This could be a "sigma effect" (see Chapter V), or due to slippage at the wall. These two phenomena should not be confused. J. G. Oldroyd has dealt with the latter, as have many other rheologists; Southorn follows Oldroyd's treatment. Schofield and Scott Blair published many papers on the "sigma effect" in the early thirties. Under the conditions of shear found in small capillaries it is very likely that non-rigid lutoids would be distorted, much as but more than has been suggested for blood corpuscles (see Chapter V).

These complex phenomena, especially shear thinning effects, have one important practical consequence. They enable hydrometer readings to some extent to be manipulated, with obvious results (desirable or undesirable according to the point of view) on the selling price of the latex, since hydrometer readings are often used as an index of rubber content. For those who wish to follow up the subject a good introductory paper is one by Southorn [22], but it is to be hoped the matter will be taken up in good faith.

Flow from latex tubes

The phenomenon of rubber latex flow from opened tubes is a complex one; a good account is given by Frey-Wyssling [16]. There are two main effects which have to be taken into account. The first arises from the fact that the tubes are somewhat elastic. When opened they therefore contract and their change in volume corresponds to the exudation. The second aspect turns on the fact that a pressure gradient is maintained partly by the supply of fluid to regions near the cut from other regions progressively further away, and partly by the intake of water consequent on the fall of the turgor pressure. The problem is in fact an extremely complex one, with tubes shrinking in diameter with time, sap falling in concentration, and the contributing length continually extending. In addition the open ends of the tubes tend to become clogged with coagulated latex. It is not surprising therefore that a good deal more work needs to be done before the process is properly understood.

Artificial enhancement of flow

One interesting discovery, as yet little understood, has been that certain chemical substances applied to the bark of rubber trees cause a considerable increase in flow. Among the earlier substances to be tried were the synthetic growth hormones 2,4-dichlorophenoxyacetic acid and the 2,4,5-trichloro-compound, and these achieved considerable success. More recently "ethrel", a synthetic substance which liberates ethylene in the presence of water, has proved remarkably effective, increasing the exudation by 60% or more (Moir [17]). It is now used regularly in many rubber plantations. Its mode of operation is unknown; but earlier work had shown that cessation of flow is connected with the plugging of the cut ends of the latex vessels with coagulated rubber; the influence of the ethylene set free may therefore be associated with an interference with this phenomenon.

The rheology of latex, at the physiological level, still remains a subject of which little is known.

The cell wall

One of the distinguishing features of plant cells (as opposed to animal cells) is the possession of a more-or-less rigid cell wall. As is well known, this is usually composed in the main of cellulose, the most naturally abundant of all organic compounds. It has been found that the long chain molecules of cellulose aggregate together in a crystalline fashion to form "micellae". These are about 50–60 Å in diameter, and they aggregate in turn in a less regular crystalline fashion to form what are often called "microfibrils" of about 200–300 Å diameter. These microfibrils are of considerable length and may contain, say, 25 micellae in their cross-section comprising perhaps 2500 elementary molecular chains. It is of these microfibrils that the cellulose framework of cell walls, multiform in its detailed architecture as it proves to be, seems invariably to be woven.

This is not the place for a detailed dissertation on the cell wall and its structure, but it should be obvious at once that many topics of rheological interest are to be found here. For instance, the newly-formed cell is surrounded from the start by a wall; how is it to grow? What processes of deformation and flow in the microfibrils lacing the wall together allow the necessary extension in area? Again, the young cell will probably change its shape, perhaps growing out into a long thin fibre, and in doing so intruding its ends between other cells. Does its surface, or at least part of it, have to "glide" over the surfaces of other cells in this process? Plant cells are invariably subject to high internal hydrostatic pressures—a most important source of the rigidity of the plant body. How does the wall respond to this pressure? What is the relation between stress and strain; does it deform plastically? No doubt the "flow" of cell wall material is not rapid, by usual standards; but it constitutes a point of obvious rheological interest nevertheless.

It may be said at once that many of the questions of importance have probably not even been posed yet in strictly rheological terms. The most that can be done therefore is to describe in outline the cell wall situation as an introduction and hope that it will prove suggestive. Further information can be found in a review by Frey-Wyssling [16] and in the splendid monograph of Roelofsen [18].

The cellulose microfibrils from which the wall is spun or woven in various ways have as two of their main characteristics great tensile strength and inextensibility. They can in fact be thought of almost as microscopically fine steel wires. Accompanying them as major constituents of the wall are amorphous hydrophilic substances such as pectins, hemicellulose and proteins, with incrusting substances such as lignin often appearing later. Thus the young wall can be envisaged as an exceedingly fine basket work of strong but flexible fibres saturated with a slimy material which fills up all the interstices, the two components having however a chemical affinity with one another which results in hydrogen-bonding, for instance. With such a structure it is not surprising that the rheological behaviour is complex.

The primary wall

Probing further, it should be noticed that from the beginning cell walls are separated from each other by a layer of amorphous material—the "middle lamella"—composed principally of pectins. This layer persists throughout the life of the cell, though no doubt it is subject to change. Next to this layer arises what is called the primary wall. This dominates the cell while it is in the process of extension growth, and no doubt its structure reflects this fact. The primary wall is composed of the usual microfibrils laid down in a more-or-less random ("dispersed") orientation and interwoven loosely with one another (Fig. XII-14). During the stage when the middle lamella and the primary wall alone enclose the cell the latter is able to grow, i.e. to increase in surface area. The interesting problem is how the microfibrils accommodate themselves during this enlargement; for instance, do they "slip" (where they cross) under the influence of turgor pressure, and continue to grow at their ends? or do they rupture mechanically or enzymatically, and have a new section intercalated? This is a classical problem, and it cannot yet be said to have been solved. Probably the best hypothesis is the "multinet" theory of Roelofsen [18]. This supposes that the loosely-woven microfibrils of the primary wall are orientated on formation in the direction of the maximum tensile stress (i.e. more-or-less perpendicular to the axis). Cells grow (for instance at the root or stem apex) mainly in an axial direction. This growth tends to pull the roughly transverse fibrils into a more axial alignment. While this is taking place the tenuous layer of microfibrils is being continuously added to on the inner face by new microfibrils, woven-in at first in a direction roughly transverse to the axis but subject to progressive axial realignment in their turn as the cell continues to grow in length. Thus at any stage the innermost fibrils of the thin, loose wall are more nearly transverse, the outermost more nearly longitudinal, with a continuous gradation between (Fig. XII-15). Owing however to the thin nature of the primary wall, and to the fact that the fibrils interweave, this pattern is not all that obvious. This "multinet" theory has found fairly wide acceptance. Of interest is the fact that it implies that the microfibrils can slide over each other, as Fig. XII-15 makes plain. It is not difficult to appreciate that such a wall has no well-defined mechanical properties; mechanical stress will cause a complex response compounded of elastic extension and internal rearrangement of a kind hardly described adequately as "plastic deformation".

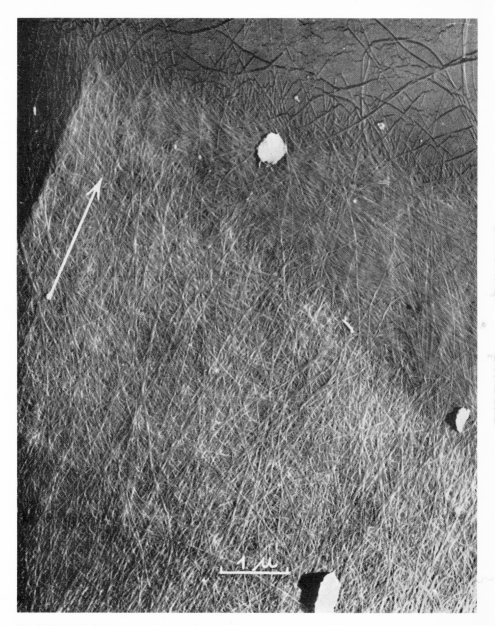

Fig. XII (14). Cellulose structure on the outer and the inner surface of the cell-wall of *Spirogyra setiformis*. Arrow indicates cell axis. Reproduced by courtesy from: Vasumati, J.L., Roelofsen, P.A. and Spit, B.J., Acta Bot. Neerl., 11: 225–227 (1962).

192

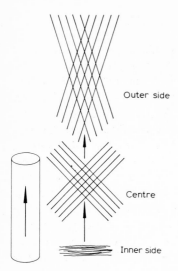

Fig. XII (15). Schema of the way in which, according to the theory of multinet growth, the orientation of the microfibrils in the successive layers of the primary wall originates. Reproduced by courtesy from: Roelofsen, P.A. and Houwink, A.L., Acta Bot. Neerl., 2: 218 (1953).

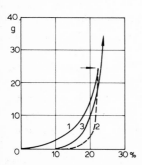

Fig. XII (16). Load-extension curve of primary cell wall (epiderm cells of *Avena* coleoptile). Load in g, extension in %. 1, first strain; 2, release; 3, repeated strain. Reproduced by courtesy from: Steyn, Protoplasma, 19: 78 (1933).

An experiment on a typical primary cell wall illustrates this (Fig. XII-16). An interesting theoretical discussion of multinet growth which takes into account such factors as fibril orientation and the viscosity of the matrix, and shows how these are related to the resulting shape of the cell has been given by Probine and Barber [19].

The secondary wall

Once extension growth of all, or part of a cell wall has ceased, the so-called secondary wall is laid down by apposition on to the primary wall. It often, in time, completely overshadows the former in thickness. In place of the more-or-less random orientation of the fibrils there appears a parallel texture with the direction running helically round the elongated cell. At one extreme (some plant fibres) the helix is so steep as to be practically axial; at the other (annular xylem vessels) the direction appears quite transverse. Most often it is in-between, an obvious helix. The secondary wall may be several-layered, with the angle of the helix distinctly different and indeed its direction often reversed, from one layer to the next (Fig. XII-17). Clearly such a structure will give the wall distinctive mechanical properties. Where the helix angle is steep the extensibility will be low and the tensile strength high; where it is low the reverse will be the case. Especially where incrusting substances like lignin are abundant as in palm fibres there will be a good deal of irreversible plastic extension or "flow" on stretching.

Intrusive growth

Often in elongated cells part of the primary wall matures and secondary wall deposition begins before all extension growth is completed. In particular the ends of the fibres and other long cells may continue to grow and insert themselves between the mature parts of other cells. This intrusive growth raises problems of its own: does "gliding" of one wall over another take place, or is there growth of wall only at the tip, the process being regarded more like the unrolling of a carpet? The latter type of growth—at the tip only—appears to be far the commoner type, but gliding growth does seem to occur in some cases. Either way, the rheological properties of the substance of the middle lamella would seem important.

Slimes

This would seem to be the place to mention the frequent occurrence of slimes in the plant kingdom. These are polysaccharide in nature and fulfil several functions, some at least of interest to the rheologist. Many seeds, e.g. cress and flax, develop surface slimes on germination. Often free cellulose fibrils can be seen in these slimes under the electron microscope, doubtless influencing their viscosity and tenacity. Root tips, as is well-known, possess a continuously-renewed cap of cells which undergoes slimy degeneration. No

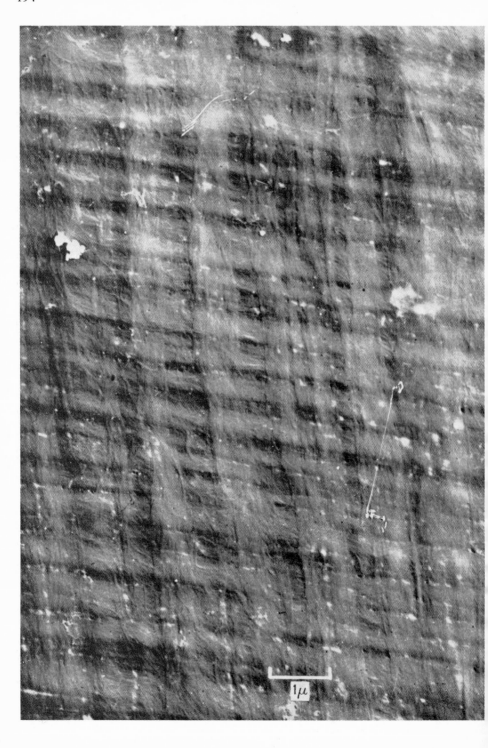

doubt one function of this cap is to assist the delicate organ to penetrate the the soil by providing some form of lubrication. How this interacts with the structure of different types of soil is a problem for some future rheologist, for soil situations can differ very widely for the young root. Again, both tenacity and deformability would seem to be advantages in the lubricating material. The root-cap cells also contain the large starch grains believed to function as gravity-perceptive organs or "statoliths". The sedimentation of these in the cells is a rheological problem belonging rather to the realm of protoplasm (Chapter III).

Of a different order are the mucilaginous substances present in the cell walls of algae, especially the larger brown marine algae like *Fucus*. These have an obvious function to fulfil in enabling the plants to withstand the violent buffeting inherent in life in an intertidal zone. A lubricating property is obviously essential here, though it is not the only property of importance; water conservation is another.

Honey

Of quite different interest is the fact that while most honeys (in the clear state) are Newtonian in their behaviour a few are markedly thixotropic. Such are the honeys of the common ling (*Calluna vulgaris*) and of the New Zealand *Leptospermum scoparium* which show marked thixotropy. Other honeys, such as those from some of the species of *Eucalyptus*, show shear thickening (see Chapter I).

The thixotropic honeys were first investigated by Scott Blair [20], and later by Pryce-Jones [21]. Analysis seemed to indicate that the thixotropic property was due to the fact that the honey contained a relatively high amount (1.3 to 1.5%) of protein. On removal of this it became normal; and if a rice protein was added to clover honey it conferred the thixotropic property. *Leptospermum* honey had the same peculiarity. The thixotropy has the commercial consequence that the honey will not flow out of the combs when the latter are rotated in the extractor; the combs have to be crushed and the wax filtered off (or sometimes, a special kind of "comb" is used). Nor will it granulate nearly so easily as other honeys. Already favoured because of its rich flavour

Fig. XII (17). *Alstonia spathulata*. Cross-banded structure in the cell wall of a tracheid. Longitudinal cell axis approximately diagonal. ×16,000. Reproduced by courtesy from: Wardrop, A.B., Aust. J. Bot., 2: 154–164 (1954) (Plate 5).

and colour, ling honey possesses the further advantage that it can be creamed in a stable fashion by stirring in small air bubbles; its thixotropy moreover encourages good table-manners.

At the other extreme are the honeys which show shear thickening in the stringy form known as "spinability". When these are stirred the viscosity rises, often quite abruptly, and to as much as ten times its original value. Shear thickening can cause difficulties in the mechanical extractor. When a glass rod is removed from these honeys it is accompanied, not by a gentle returning stream of viscous fluid at most a few inches long, but by a string which can be drawn out perhaps to several feet. This string shows semi-elastic properties, like mucus. Contrary to the case of thixotropic honey the property here seems to be due not to protein, but to dextran, which may be present to the extent of several per cent. Stringy honey can apparently be an embarrassment to bees, who may be unable to imbibe it owing to the peculiar resistance to flow. The presence of dextran is possibly connected with bacterial action.

REFERENCES

[1] Biddulph, O., in *Plant Physiology, A treatise* (Ed. F. C. Steward), Vol. II (Academic Press, New York and London, 1959).
[2] Crafts, A. S. and Crisp, C. E., *Phloem Transport in Plants* (W. H. Freeman, San Francisco, 1971).
[3] Esau, K., *Anatomy of Seed Plants* (Wiley, New York, 1960).
[4] Marshall, D. C., *Plant Physiol.*, 33: 385 (1958).
[5] Peel, A. J., *Ann. Bot.*, 29: 119 (1965).
[6] Heine, R. W., *J. expt. Bot.*, 22: 503 (1971).
[7] Zimmermann, M. H., *Biorheology*, 2: 15 (1964).
[8] Scholander, P. F., Hammel, H. T., Bradstreet, E. D. and Hemmingsen, E. A., *Science*, 148: 339 (1965).
[9] Weatherley, P. E. and Johnson, R. P. C., *Int. Rev. Cytol.*, 24: 149 (1968).
[10] Fensom, D. S. and Spanner, D. C., *Planta*, 88: 321 (1969).
[11] Churayev, N. V., Sobolev, V. D. and Zorin, Z. M., *Faraday Soc. Spec. Discussion*, 1970.
[12] Robinson, R. A. and Stokes, R. H., *Electrolyte Solutions* (Butterworths, London, 1970).
[13] MacRobbie, E. A. C., *Biol. Rev.*, 46: 429 (1971).
[14] Spanner, D. C., *Nature*, 232: 157 (1971).
[15] Milburn, J. A., *Planta*, 95: 272 (1970).
[16] Frey-Wyssling, A. (Ed.), *Deformation and Flow in Biological Systems* (North-Holland, Amsterdam, 1952).
[17] Moir, G. F. J., *Rubber Res. Inst. Malaysia Planters' Conf. Preprint* D-1, (1970).

[18] Roelofsen, P. A., "The Plant Cell Wall," in *Encyclopaedia of Plant Anatomy,* (Gebrüder Borntraeger, Berlin, 1959).

[19] Probine, M. C. and Barber, N. F., *Aust. J. biol. Sci.,* 19: 439 (1966).

[20] Scott Blair, G. W., *J. phys. Chem.,* 39: 213 (1935).

[21] Pryce-Jones, J., *Proc. Linn. Soc. London.,* 155: 1942 (1944).

[22] Southorn, W. A., *J. Rubber Res. Inst. Malaya,* 21 (4): 494 (1969).

APPENDIX
(Oxford, February 1974)

There have been unavoidable delays in the publication of this book and the author is grateful to the publishers for allowing him to add this brief appendix at the proof stage.

Not only is the study of Biorheology advancing very rapidly, but, since the writing of this book was completed, the 1st International Congress on Biorheology was held at Lyon in France, and most of the papers read there are being published (some in modified form) in the official journal of the International Society, "Biorheology". Some of these could be mentioned in the main body of the present book, but those who have found the subject of interest are advised to consult the Journal, Vols. 10 and 11 (1973, 1974) where most of the original papers are being published. In the meantime, a few more key-references should be given as a postscript to the book itself.

Chapter I. Electro-rheology

Fukada [1] and his colleagues have published a number of papers on piezo-electric phenomena first discovered in wool cellulose. Such effects were found by the Japanese workers to occur in myosin, actin and other biological materials. If an anisotropic material is sheared in an appropriate plane, an electrical polarization is produced in a direction perpendicular to the shear plane. Similar effects were found by Fukada and Hara [2] in blood vessel walls.

Chapter III. Protoplasm

Jahn and Votta [3, 4] recently published two important papers on "birefringence as an index of amoeboid movement" and on a "capillary suction test of the pressure gradient theory of amoeboid motion." They criticize some conclusions of Allen [5] and his colleagues concerning the flow of protoplasm when sucked into a capillary, and the alleged production of flow-birefringence. But the discussion is too complex to be dealt with here, and the reader is referred to the original papers.

Chapter IV. Muscle, collagen, bone, brain, hair, skin, etc.

Many interesting papers on the rheology of these systems are to be found in J. Biomechanics, especially Vol. 5, published too late to be discussed in this book (see especially Hang-Chu-Wang and Wineman on brain [6]).

A paper by Parker [7] on muscle should be noted. He measured the dynamic modulus over a wide range of frequency. Muscle is found to be a highly non-linear viscoelastic material. For a criticism of Rigby's work on collagen* see Cohen et al. [8].

An interesting unsigned article on the short- and long-term effects of boxing, by professionals, amateurs and even schoolboys is to be found in the Brit. Med. J. for 24th November 1973.

A new collection of articles on muscle biology (including rheology) is edited by Cassens [9].

Stoner et al. [10] have proposed an elastic-viscous model for muscle consisting of only four elements: two dashpots and two springs.

An excellent review article on rheology and biochemistry of muscle is to be found in "The Scientific American" for February 1974 by J.M. Murray and A. Weber (p. 59).

Chapter V. Blood (flow); plasma: serum

A review of recent theoretical work on haemorheology (with 56 references) is given by Oka [11].

Rosenblum and Warren [12] have pointed out a potential source of error in the use of rotation viscometers for blood. There can be an increase in viscosity produced by shearing. This is comparable to that found in a capillary viscometer by Scott Blair and Matchett [13].

Eastham [14] stresses the advantages of using J. Harkness' measure of viscosity of plasma in place of ESR, claiming that this test involves less risk of hepatitis for the technician. He quotes other authors who have expressed similar views.

For further reading: "Blood Flow Measurements" by V.C. Roberts, Robert Sector Publ. Ltd. London 1972.

---

*See ref. 26, p. 45

Chapter VI. Blood coagulation

As an addition to the French "Atlas" on thrombelastography, readers who know Italian, are recommended to a small book from the University of Milan by Diomede-Fresa and Fumarola [15]. Also recently published is an admirable review article by Copley [16] on tests for bleeding time and the arrest of haemorrhages.

Chapter IX. Sputum, bronchial mucus and salivas

The author is much indebted to Professor Lynne Reid and her colleagues who report on work done since this chapter was written. "In a series of chronic bronchitics with severe airways obstruction a correlation was found between the viscosity of sputum as measured with a Ferranti Shirley viscometer and the severity of the airways obstruction (Lopez-Vidriero, Charman, Keal, De Silva and Reid [17]). Although at one time N-acetyl neuraminic acid was thought to be important in determining sputum viscosity (Keal [18,19]), a stronger correlation between fucose and viscosity has now been established. While fucose and neuraminic acid are both constituents of the acid glycoprotein of epithelian mucus, serum contains virtually no fucose and has a low viscosity. It is unlikely, however, that the physico-chemical properties of fucose, a pentose sugar, can explain the close correlation with viscosity. Fucose is probably better regarded as a marker or measure of the acid glycoprotein constituent of sputum whereas neuraminic acid represents, in addition, the tissue fluid component, either transudate or exudate.

Freezing sputum has been said to alter its rheological properties (Sturgess, Palfrey and Reid [20]). Controlled experiments, at various temperatures for freezing, storing and thawing have now shown that degradation as determined with the Ferranti Shirley viscometer, is prevented by rapid freezing in liquid nitrogen and storage at $-13°C$ or below and also by both freezing and storage at $-70°C$. The temperature of thawing was found not to be critical (Chapman and Reid [21]).

Elasticity of sputum has been determined using the Weissenberg rheogoniometer, by oscillatory testing, over frequencies from 0.01 to 0.8 cps. (Mitchell-Heggs, Palfrey and Reid [22]).

Sputum was from patients with either asthma, chronic bronchitis, cystic fibrosis or bronchiectasis, and macroscopically mucoid mucopurulent or purulent.

As in the viscosity plot over this shear rate range two zones with a junctional region can be distinguished. In zone 1, over the lowest shear rates, elasticity increases slowly, changes little over a 'plateau' region and then, in zone 2 increases sharply. In contrast to the viscosity plot, the plateau does not show notching. By 0.8 cp some samples show decreasing elasticity. Although variance between samples is wide, there is no level of elasticity characteristic of each disease or of one macroscopic appearance.

Elasticity and viscosity are correlated, most significantly at the lowest shear rates. Asthmatic and bronchiectatic sputa resemble each other in that this correlation is still significant at higher shear rates, cystis fibrosis and chronic bronchitic sputa in that it is not. Since in these last two mucous gland hypertrophy is present, it may be that the sputum has a higher mucus component."

Chapter X. Miscellaneae

My attention has been drawn to a paper concerning the rubber-like elasticity of a substance called "surume" derived from Japanese cuttle-fish. See Kishimoto and Fujita [23].

Several recent papers have been published in J. Biomechanics on mucuproteins from urine which will be of marginal interest to rheologists.

Raju and Devonathan [24] have published a highly mathematical study of peristaltic matian in non-Newtonian (especially power-equation) systems. It is assumed that the peristaltic action is sinusoidal.

Relating to Surface Rheology, Copley and King [25] describe a modification of the rheogoniometer suitable for measuring the properties of surface layers, using this to study the effects of γ globulin and β lipoprotein in the rheology of fibrinogen.

Chapter XI. Psycho-rheology

Many interesting papers continue to be published in the Journal of Texture Studies and a new Journal has recently appeared: "Chemical Senses and Flavor" (D. Reidel Publ. Co.) which covers the tactile and visual properties of materials.

APPENDIX REFERENCES

1. Ueda, H. and Fukada, E., *Jap. J. appl. Phys.*, 10: 1650 (1971), and other papers.
2. Fukada, E. and Hara, K., *J. phys. Soc. Japan*, 26: 777 (1969).
3. Jahn, T.L. and Votta, J.J., *J. Mechanochem.* Cell Motility, 1: 245 (1972).
4. Jahn, T.L. and Votta, J.J., *Science*, 177: 636 (1972).
5. Francis, D. and Allen, R.D., *J. Mechanochem. Cell Motility*, 1: 1 (1971).
6. Hang-Chu-Wang and Wineman, A.S., *J. Biomechan.*, 5: 431 (1972).
7. Parker, N.S., *Rheol. Acta*, 11: 56 (1972).

8. Cohen, R.E., Hooley, C.L. and McCrum, N.G., *Nature*, 247: 59 (1974).
9. Cassens, R.G. (Ed.), *"Muscle Biology: a Series of Advances"*, Marcel Decker, New York 1972.
10. Stoner, D., Haugh, C.G., Forrest, J.C. and Sweat, V.E., *J. Texture Studies*, 4: 483 (1974).
11. Oka, S., *Advan. in Biophys.*, 3: 97 (1972).
12. Rosenblum, W.J. and Warren, E.W., *Biorheology*, 10: 43 (1973).
13. Scott Blair, G.W. and Matchett, R.W., *J. Neurol. Neurosurg. Psychiatr.*, 35: 730 (1972).
14. Eastham, E.D., *Brit. Med. J.*, 4: 612 (1973).
15. Diomede-Fresa, V. and Fumarola, D., *Edizioni Premio Janassini*, Milan, 1956.
16. Copley, A.L., *Thrombosis Res.*, 4: 1 (1974).
17. Lopez Vidriero, M.T., Charma, J., Keal, E., De Silva, D.J. and Reid, L., *Thorax*, 28: 401 (1973).
18. Keal, E., *Postgraduate Medical Journal*, 47: 171 (1971).
19. Keal, E., *in Pathologie et Nosologie de la Bronchite Chronique, Colloque de Lyon*, 22/23 Sept. 1970, p. 15. Edited by L'Association Médicale et Scientifique de Lutte contre la Bronchite Chronique, 1971.
20. Sturges, J., Palfrey, J. and Reid, L., *Rheol. Acta*, 10: 36 (1971).
21. Charman, J., Reid, L., *Biorheology*, 10: 295 (1973).
22. Mitchell-Heggs, P., Reid, L., Palfrey, J., *Elasticity of Sputum* (in process of publication), 1974.
23. Kishimoto, A. and Fujita, M., *Bull. Jap. Soc. Sci. Fish.*, 22: 293 (1956).
24. Raju, K.K. and Devonathan, R., *Rheol. Acta*, 11: 170 (1972).
25. Copley, A.L. and King, R.G., *Thrombosis. Res.*, 4: 193 (1974).

AUTHOR INDEX

206

SUBJECT INDEX